STEWART-THORNYCROFT STEAM WAGONS.

5-ton End Tipping Wagon.

Embody the result of 9 years' practical experience. . . . . . Write for price and particulars, stating requirements.

D. STEWART & CO. (1902) LTD.
London Road Ironworks, GLASGOW.

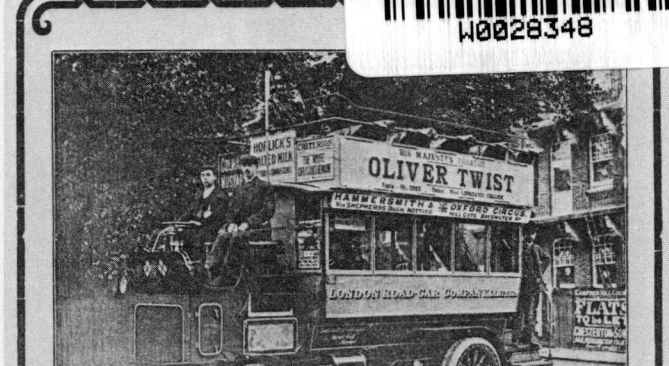

The third year of successful Public Service in England.

These Cars have no Change Gears and no Clutch.

They have Direct Drive at all Speeds, hence the characteristic smooth and swift movement.

Paraffin Fuel is used.

CLARKSON LTD.,
Chelmsford.

The LONDON COUNTY COUNCIL

have just favoured us with an order for FIVE of our Standard Steam Wagons.

R. C. BAULY, ESQ., of 131a, BOW ROAD, E., writes, January 11th, 1907:—
"I have much pleasure in sending you herewith an order for 4 more of your Standard Pattern 5-ton Wagons, which makes a total of NINETEEN Wagons I have ordered from you during the last two years, and I am glad to say they are giving every satisfaction."

FODENS LIMITED, SANDBACH.

ROBEY & CO., Ltd., GLOBE WORKS, LINCOLN, ENGLAND.
Patent Steam Wagons for all Countries.

—— Prices and Descriptive Catalogue on Application. ——
STAND No. 236, ROYAL SHOW, LINCOLN, JUNE 25-29.

WILLIAM ALLCHIN, Ltd., Globe Works, NORTHAMPTON.

Manufacturer of

HIGH-CLASS STEAM MOTOR WAGONS

for BREWERS, MILLERS, CONTRACTORS, Etc.

Having a Tank capacity for a twenty miles journey. 2 Driving Chains in dust-proof cases. Loco Type Boiler. Engine works in Oil Bath. Accessibility to all parts.

—— ILLUSTRATED CATALOGUE SENT ON APPLICATION. ——

THE "GORTON" LORRY.

BEYER, PEACOCK & CO., Ltd.,
Gorton Foundry, MANCHESTER.
Established 1854.

34, Victoria St., Westminster, S.W.

Telegrams: "Loco, Gorton."
Telephone: 255 Manchester.

Telegrams: "Folgoro, London."
Telephones: 746 and 747 Westminster.

Highest Class Steam Motor Vehicles,
FOR ALL PURPOSES, OF THE SIMPLEST POSSIBLE DESIGN.

PREFACE

THE attraction of contributing to *OLD MOTOR*'s Kaleidoscope series was that, like conversation, such a book could range at will without having to be predictable or over-technical. The period covered by the book extends over about a hundred years but the zenith of steam wagon usage in Great Britain was from about 1913 to 1925. The peak number of wagons in use just topped the 9000 mark around 1925, a considerable number, but a drop in the ocean compared with the total of goods vehicles that year which was around 225,000. This is to say, four per cent of the total were steamers but this small numerical fraction took a very considerable part in heavy and rough service. The book is mainly, but not exclusively, about steam wagons in Britain. The numbers of wagons used elsewhere in Europe probably never reached four figures but the French were early experimenters and the Germans very recent ones. American builders were early in the field but soon faded. Interest in steam commercials has never expired in America but nor, on the other hand, has there been a working steam wagon, in the truly commercial sense, for over fifty years.

Steam wagon makers, on the whole, did not deal in horsepower ratings. They stated a nominal load capacity - which operators exceeded by up to a hundred per cent - and invited their customers to accept, or to determine by experience, that the designer knew how to put enough punch into his wagon for it to do what was required of it. As a rough guide to power output, the bhp figures for Garrett overtypes may be of some use. The three tonner has an output of 30 bhp at working revs, and the five tonner thirty-five. The thermal reserve and powers of recovery of the boiler in ordinary service were, however, very important elements of practical success and were properties not much brought out by sustained tests on the brake. I have, therefore, generally written maker's described carrying capacity rather than horsepowers in description.

A great many people have helped with the preparation of this book. I am indebted particularly to Dr Richard Roosen for comments on the Henschel wagons and for the two pictures of them in Chapter 4. Much help has been given by Nick Georgano of the National Motor Museum at Beaulieu and by Messrs John Creasey and David Phillips of the Museum of English Rural Life at Reading University. Friends and acquaintances who have kindly lent photographs include: Philip Bradley, Leslie Burberry, Tony Coombes, J K Ellwood, Barry Finch, the late Bill Hughes, Arthur Ingram, Charles Lloyd, Alan Martin, Ken Mills, Dr John Middlemiss, Tom Paisley, Derek Stoyel and D Waddingham. W & J Glossop Ltd kindly lent photographs of two of their Atkinson tar sprayers and Richard Garrett Engineering Ltd lent pictures of their products.

To all who have contributed, whether mentioned or not, I extend hearty thanks.

THE AUTHOR

THE author grew up in a village where steam engines were plentiful. Two of his father's friends owned a steam threshing set powered by a Fowler single cylinder traction engine. In fact, they bought a new one in 1934 - the last Fowler agricultural engine made.

His training in steam, however, took place at the Ashford Works (Kent) of the former Southern Railway, but he abandoned the railway for a career in the construction industry, in which he has spent the greater part of his working life.

He began writing in the winter of 1939/40. Most of his early works, including his first book, were concerned with railway subjects but he has since written six books on road steam (and has another in preparation) besides numerous articles.

THE YEARS BEFORE 1900

Until the partial relief afforded by the 1896 Act, mechanical road vehicles in Britain were looked upon as road locomotives - that is, essentially, as traction engines. Because of this legislative emphasis and of the relatively extended use of traction engines in agriculture such development as there was of steam haulage on roads was based upon the use of traction engines.

1 Amedée Bollée was a bell-founder in Le Mans, Sarthe [France], who became interested in mechanical traction in the seventies, designing and building a limited series of road haulage tractors in which a heavy rigid chassis carried a vertical boiler with central uptake and Field tubes with a front mounted vertical two cylinder engine which drove a transverse countershaft through bevel gears. Final drive was by chain. In *Marie Anne* of 1879 he was treading very close to the steam wagon threshold for she was equipped with a separate tender, power driven by a counter shaft linked to the primary countershaft by a prop shaft, bevel gears and universal joints.

Across the Channel the situation was rather different. Although French roads were no better - and possibly worse in some respects - than those in Britain, the restraints on mechanical transport were practical and commercial rather than legislative. Moreover, steam power in use on farms was provided in the main by fixed engines and portables rather than by tractions so that the direction of thought on the subject was less channelled by precedent.

French experimenters with steam wagons being thus freer from inhibiting factors began sooner than their British counterparts and with lighter vehicles, rooted more in steam car and carriage prototypes than in heavy haulage vehicles. For instance, the Le Blant vehicle built for 'La Jardinière' of Paris, in 1894, was a light, smart tradesman's equipage rather than a tool for the heavy haulier, though French makers were producing steam lorries intended for heavy work before the end of the last century.

On this side of the Channel a few makers took advantage of the reliefs afforded them in 1896 to build light commercial steamers, often with flash or semi-flash boilers but the greater emphasis was upon rather heavier compound undertypes using, in the main, fire tube vertical boilers, some of them drawing their initial inspiration at first or second-hand from the French, though evolution soon produced characteristics in the British product not seen elsewhere.

There was, however, a second line derived from the strong traction engine tradition here and pioneered by the steam tractor designed by Philip Parmiter of Tisbury, Wilts, which was arranged to carry, when desired to do so, a large transport box on its own frame. This was the progenitor of the Mann steam carts and, in all but the strictest sense, of the overtype steam wagons which played a large part in establishing steam lorries as a viable tool for the British trader. Fodens produced their overtype before the new century, soon to be followed by Straker and, at a respectful distance, by some of the other traction engine makers.

2 In 1880 Bollée built *La Nouvelle*, a steam coach big enough to carry a dozen people. It must have been a robust, if over-heavy vehicle, for it was still in active use in 1897 and took part in the Paris-Bordeaux trial of that year. It is difficult to know if he intended it to be a private carriage or a public service vehicle. It was as big as many of the horse-drawn station buses of the period but no record survives of its being used in this way. Bollée did not pursue steam commercial vehicle development further. His son turned to steam cars and, soon afterwards, to the internal combustion engine.

3 Whilst Bollée was experimenting in France Philip Parmiter, a clever young Wiltshire agricultural engineer, had the idea of making a light [by traction engine standards] steam tractor, capable of carrying a transport box on its own frame. The rear of his tractor was carried on a single broad wheel or roller which spread the load and prevented rutting of the surface, but led him into trouble with the absence of differential action.

10 Though Parmiter had his idea in the eighties, it was not until he met James Mann of Mann & Charlesworth, newly established as traction engine and steam roller makers in Leeds, that he found a manufacturer interested in joining with him in putting his invention on the market. The example shown was built by Mann in 1898 and was followed by others, but the single broad wheel was soon abandoned in favour of a pair of wheels with differential action. Known as the Mann Steam Cart - Parmiter's part was soon forgotten - it was in production for nearly twenty years but its inventor had the last word. His firm still flourishes but Mann's went into liquidation in 1928.

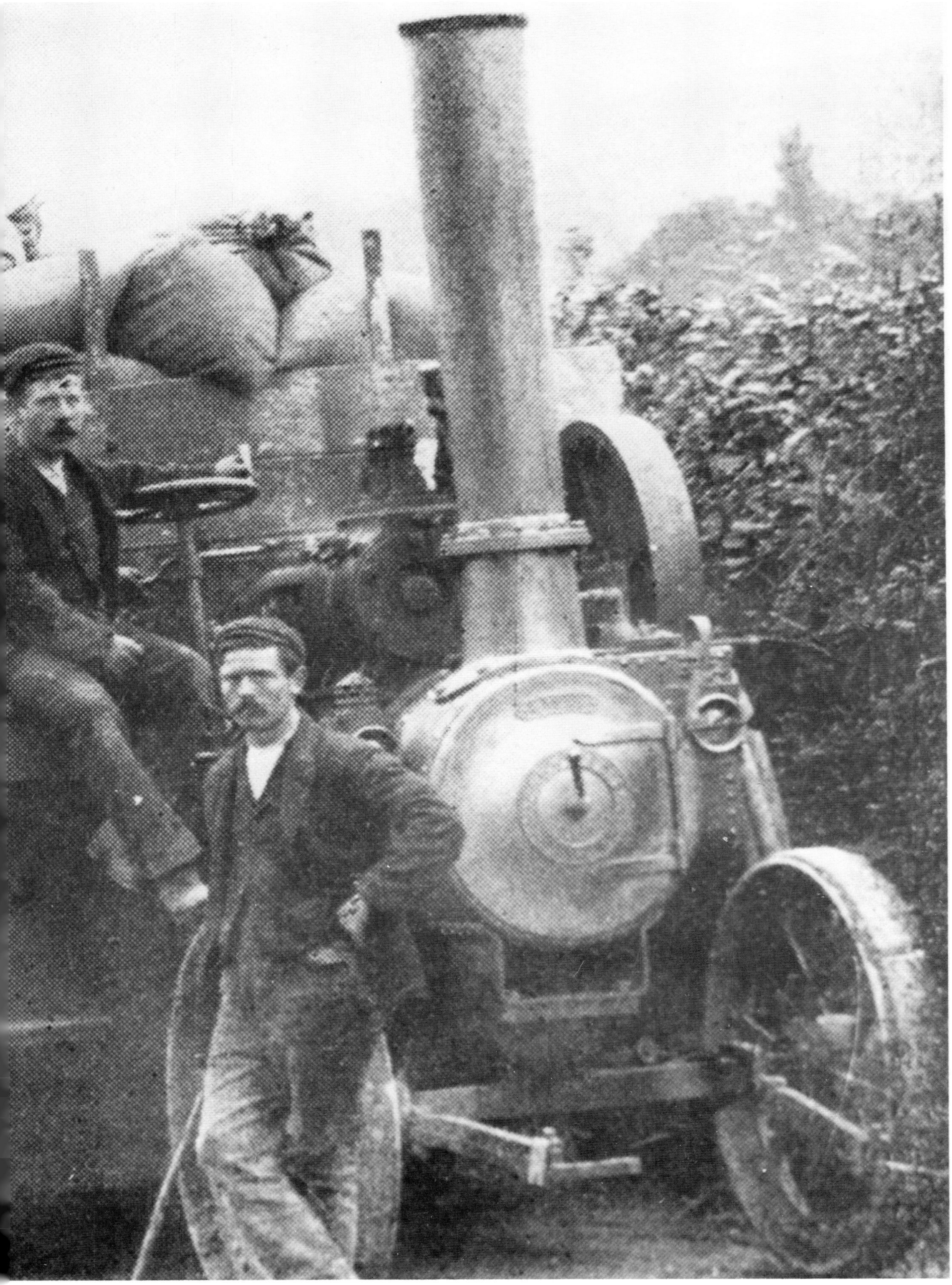

4 An inventive Maconnais hatter, Scotte, who lived at Epernay, Marne, began experimenting with steam vehicles in the late eighties and about 1892 was building some which were on the borderline between cars and buses. Scotte was his adopted surname. His patronym actually was Crotte [English - *dung* or worse] which, understandably, he found to be the butt of too many pleasantries from his acquaintances. In the 1894 Paris Rouen trial he entered a 12 seater [14 seater, according to some reporters] dubbed 'the four poster'. Following Bollée, he used a vertical boiler with Field tubes and a vertical two cylinder engine with single primary chain to the countershaft, which housed the differential, and from there to the rear wheels twin chains. The vehicle had boiler trouble - probably a tube - on Gaillon Hill, but was awarded a 500 franc prize for the best non-finisher.

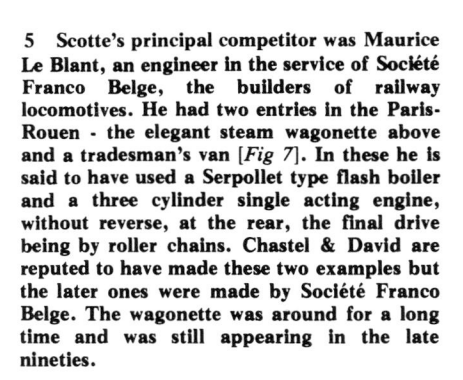

5 Scotte's principal competitor was Maurice Le Blant, an engineer in the service of Société Franco Belge, the builders of railway locomotives. He had two entries in the Paris-Rouen - the elegant steam wagonette above and a tradesman's van [*Fig 7*]. In these he is said to have used a Serpollet type flash boiler and a three cylinder single acting engine, without reverse, at the rear, the final drive being by roller chains. Chastel & David are reputed to have made these two examples but the later ones were made by Société Franco Belge. The wagonette was around for a long time and was still appearing in the late nineties.

6 Le Blant's second entry in the 1894 trial was a smart equipage for a Paris tradesman which probably earned him valuable publicity but otherwise could scarcely have earned its keep. By 1894 Le Blant had begun to experiment with steam tractors for which he became famous - or notorious, according to point of view. With these he intended to work trains of two or three four-wheeled trailers, carrying passengers and light parcels along rural roads, in much the same way as the French roadside steam tramways. Numerous schemes were tried in the next ten years using Scotte, Le Blant and De Dion steamers, but all came, in the end, to nothing, mostly because each was set up in isolation, without relief vehicles, and with no proper facilities or staff for maintenance, so that timekeeping and reliability became a joke.

7   The partnership of Count Albert de Dion with Georges Bouton and the latter's brother-in-law, the pessimistic blacksmith Trèpardoux, has always seemed one of the more unlikely formulae for a successful business. Trèpardoux did not stay the course but Bouton and the Count set up a firm that went on until 1950. At first they concentrated on avant-train tractors, designed to take the place of the fore-carriage of a horse-drawn vehicle - an idea which attracted many designers for a decade or so - but went on to build rigid commercials of which their 1896 bus, above, is an example. The vertical boilers of De Dion-Bouton vehicles had a central combustion area surrounded by a water and steam jacket, crossed by numerous short fire tubes and surrounded by an outer jacket which formed the smokebox. These were effective steamers, producing very dry steam, though heavy on tubes, and went on being made for about forty years. Abbotts, the Newark boiler-makers, made one for a Liverpool firm as late as 1932. The totally enclosed engine was underfloor and drive was by cardan shaft.

8   Serpollet was one of the flash-steam school. His earlier vehicles, mostly cars, used his patented form of 'squashed tube' for the steam generator but later he went over to plain tube. In the mid-nineties he was producing self-contained double decked steam trams for the Paris tramways. Serpollet cars were set to work in June 1896 on the line from the Madeleine to Asnières, cutting the journey time, compared with horses, from 50 to 35 minutes. The same year the General Omnibus Company of Paris offered a prize for the best design of mechanically propelled bus to replace their current standard model of double decked horsebus. Serpollet did not compete directly but two years later produced this bus which took part in the heavy vehicle trials at Versailles. Unlike the other vehicles shown so far, it burned paraffin.

9   The only steam bus to result directly from the General Omnibus competition in Paris was the Weidnecht. Like De Dion and Bouton, Weidnecht began with avant-train tractors and produced his double decker for the competition. With front wheel drive and rear steering, it must have been interesting, to say the least, to drive. The coke-fired vertical water tube boiler served a twin cylinder engine with a single eccentric valve gear, countershaft differential and twin chain final drive. The General did not adopt it. In June 1898 a similar single decker [or the same bus rebuilt?] was brought to London for demonstration by H J Lawson as part of his current steam-bus humbug, more concerned with gathering in investors' money than putting buses on the streets, but it was too wide for use in London and, as far as is known, neither it nor the double decker ran in revenue-earning service.

11 The passage of the 1896 Act encouraged British makers to put into practice the ideas they had been mulling over mentally whilst still chafing in their legislative fetters. At Leyland, James Summer had made a not very successful wagon in 1880/82 before turning to making his famous steam lawn mowers. In 1896, joined by the Spurrier family, the firm once again turned to steam vehicles and produced the chain drive two ton van which took a silver medal in the 1897 trials at Crewe organised by *The Engineer*. The vertical compound engine took steam from a paraffin-fired vertical fire tube boiler. In many ways reminiscent of Le Blant's van for La Jardiniere three years before, it had features not used by Le Blant, notably three road speeds controlled by friction clutches. A four tonner developed from this design was awarded first prize in the Liverpool Self Propelled Traffic Association trials in 1899 but the company marked the new century with an entirely new design.

13 H A House was largely responsible for the design of the Lifu road vehicles made by the Liquid Fuel Engineering Co Ltd, East Cowes, Isle of Wight, who began making their paraffin-fired twin drum water tube boilers for use afloat in launches and yachts. His first commercial, a one tonner, entered in the 1897 Crewe trial, had one of these patent boilers admidships and each rear wheel driven separately by its own tandem compound engine through an internal gear on the wheel. The rate of fuel feed was regulated by a needle valve activated by boiler pressure. This was an almost total failure and the next year a redesigned version appeared with the boiler at the front, a piston valved horizontal compound engine and an all-gear drive. Lifu boilers were in copper throughout, making them expensive in first cost and the maintenance of a Lifu was expensive, even by early wagon standards. In 1900 the works was closed and sold, but production continued for a further period, first by the use of sub-contractors and, later, licensees, notably Belhaven. The picture shows a Lifu taking part in the 1898 Liverpool Self-Propelled Traffic Association Trial.

12 Jesse Ellis ran a successful steam contracting and engineering business in Maidstone, Kent, and followed with admiration the work being done in France on steam vehicles. The new freedom of 1896 tempted him into wagon design and he patented this 'Colonial buck-wagon' in 1897, though actual production was hindered by the outbreak of typhoid in Maidstone in the summer of 1897. A large vertical boiler with Field tubes was placed at the rear of the wagon and the engine was a three cylinder radial placed at the extreme rear, geared to a countershaft and differential from which twin chains drove very large sprockets bolted to the rear wheels. Steering was done from the vulnerable open seat at the front but the driving position was at the rear. A second version had a fixed cab with a roof mounted condenser and a third, further modified, example, was made. The buck-wagon was designed for use in South Africa and it may have got there, but there were no repeat orders.

14 James Beach & Co were pioneer garage proprietors in St James' Street, Taunton, Somerset, selling and repairing Daimlers and other cars of the period and hiring out a Daimler wagonette for local social functions. In 1899 they built the steam van shown, which weighed just under a ton and a half and used a paraffin-fired vertical boiler and a single cylinder engine. Designed speeds were 5 and 10 mph respectively and gear changes were made by means of a friction clutch.

15 The 1899 Liverpool trials attracted efforts from several notable designers, including an early model by Thomas Clarkson, soon to earn fame with his steam buses. Using a modified form of the Merryweather fire engine boiler, generating steam at 200 psi, he kept his steam pipes short by placing his vertical compound engine in the back of the cab. Boldly he ran the engine at 600 rpm, rather more than double the engine speeds used by his fellow competitors, and fitted a cab roof condenser, but the design failed to win customers and was soon changed.

16 Julius Harvey & Co of 11 Queen Victoria Street, London, were agents for Lifu vehicles and in December 1897, to publicise the vans and to encourage purchasers, they entered into an experimental six week contract with the Post Office to run a van nightly from Mount Pleasant Sorting Office in London to Redhill, Surrey, and back [a round trip of about 50 miles]. Beginning on 16th December the van left Mount Pleasant each evening at 10.30 pm, being timed to arrive at Redhill at 1.42 am, and to arrive back in London at 4.45 am. The best run was on 30th December when it arrived back an hour early. Though it was a notable pioneer effort - particularly on the part of the maintenance department - the Post Office remained coy about using steamers except on a contract hire basis for short periods.

17 Sir John Thornycroft's claim to fame rests at least as much upon the ships he built as upon the firm of vehicle manufacturers he founded. His first project, produced more as an amusement than a commercial project, was for the much illustrated one ton steam van which, fortunately, still exists, entered in the 1896 Liverpool trials, but the power unit was a bought-in launch engine and he soon set about an entirely fresh design which was probably the first successful steam lorry. In his water tube boiler, the upper and lower drums formed horizontal rings linked by sixty-eight water tubes arranged in three concentric circles, all enclosed in a dry back case and working at 200 psi, later stepped up to 225 psi. The engine was an enclosed undertype compound with radial valve gear and the all-gear drive used bronze double helical gears between the countershaft and rear axle. Thornycrofts were very well made but were dear and needed a very careful driver as water level was critical. Consequently they attracted only a high class of client, able to afford good drivers and maintenance staff. Thornycrofts dating from the last century saw almost twenty years in service, a remarkable achievement for such an early wagon.

Wheels were not the least of the problems which beset the pioneer builders of steam wagons. The traditional wooden wheel was designed to withstand dead-load rather than torque, ie, the propulsive force - horse or traction engine - was applied to the body of the vehicle and the basic function of the wheels was to hold it up. In a steam wagon the tractive force was applied to the wheel which had to transmit this force to the body and load, a function which the slender, and often elegant, wheelwright's wheels of the early steam lorries and buses were not able to sustain for very long.

The paramount problem of the early compound undertypes was steam supply. Though some of the early boilers were frankly oddities, the problem with most was simply lack of flexibility. To meet the legislative weight restrictions - and to keep down prices - the size of boilers was reduced as far as possible and consequently designers were trapped into equating the average steam requirements of the engine with the peak requirements. Many boilers which were designed to give a steady output equivalent to the average consumption by the engine were totally winded by peak demands - often on hills or starting from cold. In later wagons vertical fire tube boilers largely yielded place to cross water tube boilers which were quicker to respond to sudden demands or to locomotive boilers which could be given a greater thermal reserve.

The steam passage from boiler to cylinder block was very short and direct in an overtype wagon but elongated and exposed in most undertypes. Lagging gave some degree of mitigation of the latter problem but in all undertypes the steam underwent appreciable cooling between boiler and cylinders. In an unsuperheated wagon this meant that the steam arrived at the engine rather wet and in some degree reduced in pressure, lowering the power output from a given weight of steam and wasting the precious steam generating capabilities of the boiler. The effective superheating of the steam as it left the boiler overcame the problem in terms of performance - though it did not, of course, eliminate the thermal waste - but, except in flash steam wagons, the benefits of superheat were not appreciated in the period under review in this chapter.

Many experimenters either underestimated the problems they faced or over-estimated their capacity to solve them, producing, as a result, what the late Worby Beaumont described as elaborate scrap, but because of their collective exertions, by the end of the period considerable progress had been made in defining and resolving the problems of wagon design and it becomes us to doff our caps respectfully, rather than, with the inestimable advantage of hindsight, to scoff at what they achieved.

18 T Coulthard & Co were an old established firm of millwrights and engineers who spent some five years experimenting with the wagon shown before launching it or a sister wagon as a production model at the 1899 Liverpool trial. Like many early wagons it burned paraffin but was unusual in having a triple expansion non-reversible compound engine with piston valves, the reverse and three forward speeds being provided in the gears though the primary drive was by single chain and the final drive by twin chain. This design was over-ambitious but the succeeding solid fired wagon, which has a vertical compound engine, took a Gold Medal in 1901. Coulthard's wagon interests were joined to Leyland in 1907.

19 In St Lawrence, Newcastle-Upon-Tyne, Toward & Co carried on a partnership as boiler makers and engineers. Meek, one of the partners, was keenly interested in steam cars and in 1897 collaborated with Atkinson & Phillipson, local coach builders, in making a steam wagonette. They had patented a design of inclined vertical water tube boiler, rectangular on plan, which later in 1897 they used in conjunction with a duplex cylindered undertype engine, having Stephenson link motion reversing gear, in a one ton van for a Newcastle chemist. In 1899 they built the 3½ ton wagon shown for a Yorkshire iron ore mine and also built a small steam tractor, but otherwise confined their steam vehicle interest to making components for others.

20 In 1899 when Jacob Irgens of Bergen built this charming bus, his country was still part of Sweden and, with the possible exception of a somewhat hazy early vehicle said to have been made by Norrgber, the Swedish lock maker, it was possibly the only steam bus or lorry made in Scandinavia. Irgens used a Toward boiler, made in Newcastle-Upon-Tyne, and a three cylinder single acting engine. The all-gear drive worked through the front [steering axle] which must have made it unusual to drive. Steam heating was installed in the saloon for winter use.

21 Sixty years ahead of its time the steam bus of 1898, by David Martyn & Co Ltd, made at Hebburn-on-Tyne, had all its machinery under the platform. Single speed, it was driven by a twin cylinder engine with link motion reversing gear, unusual in incorporating an aluminium casing and other aluminium components. The coke fired boiler was described as a horizontal diagonal multitubular type. It is unlikely that the bus saw revenue earning service.

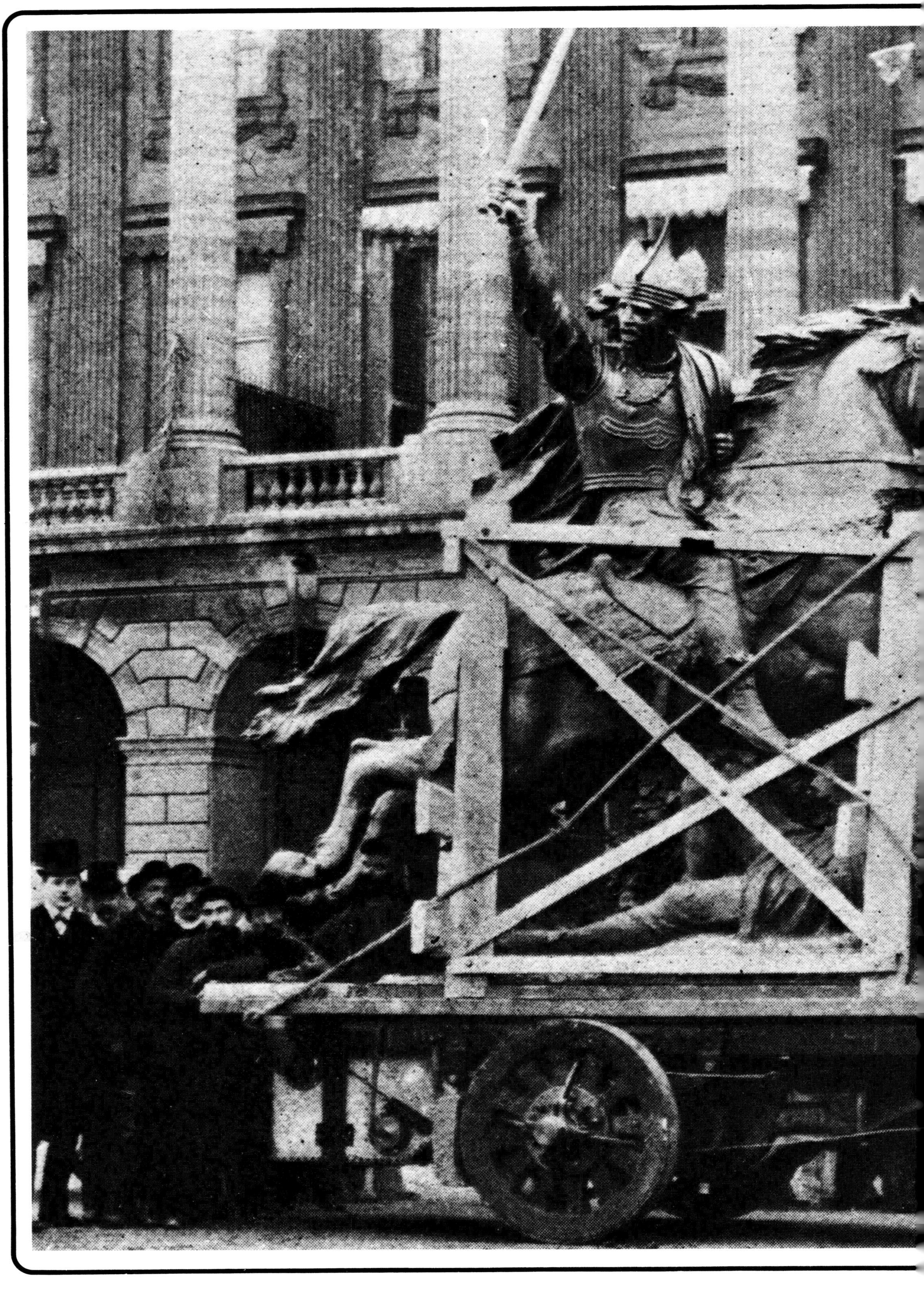

22 By the end of the last century De Dion and Bouton were turning out one heavy steamer a week from the Puteaux works and it was stated by the *Automotor Journal* in 1908 that by the end of 1901 the firm had made three hundred buses, presumably all steam driven. Five 22-seater De Dion steam buses were set to work in 1901 on a 42 mile route in Spain from Santiago to Coruna, with a 5 hour journey time. Users of their lorries included the Paris Municipality, one of whose fleet is seen here hauling a statue of Vercingétorix, the French folk hero.

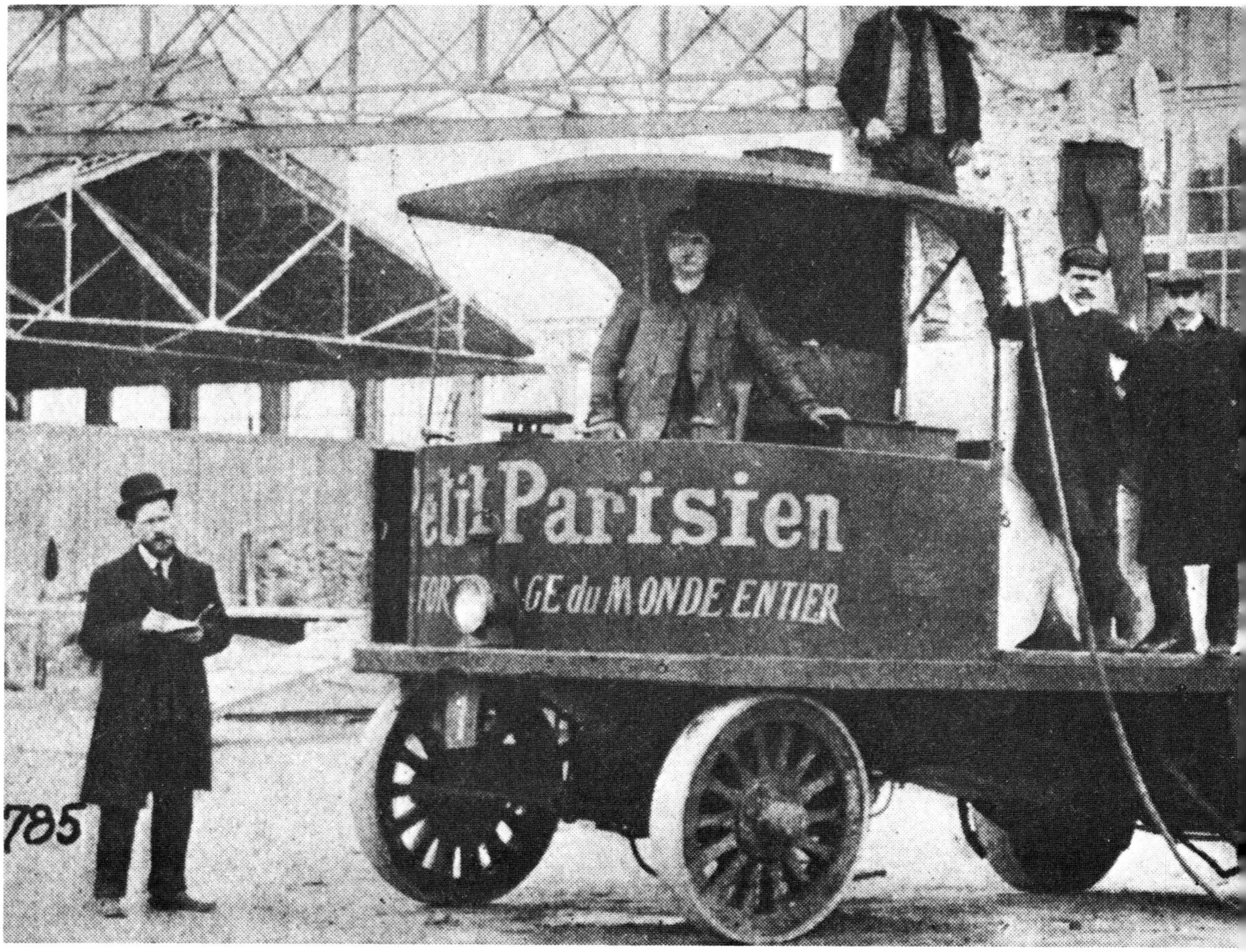
Parisien
GE du MONDE ENTIER
785

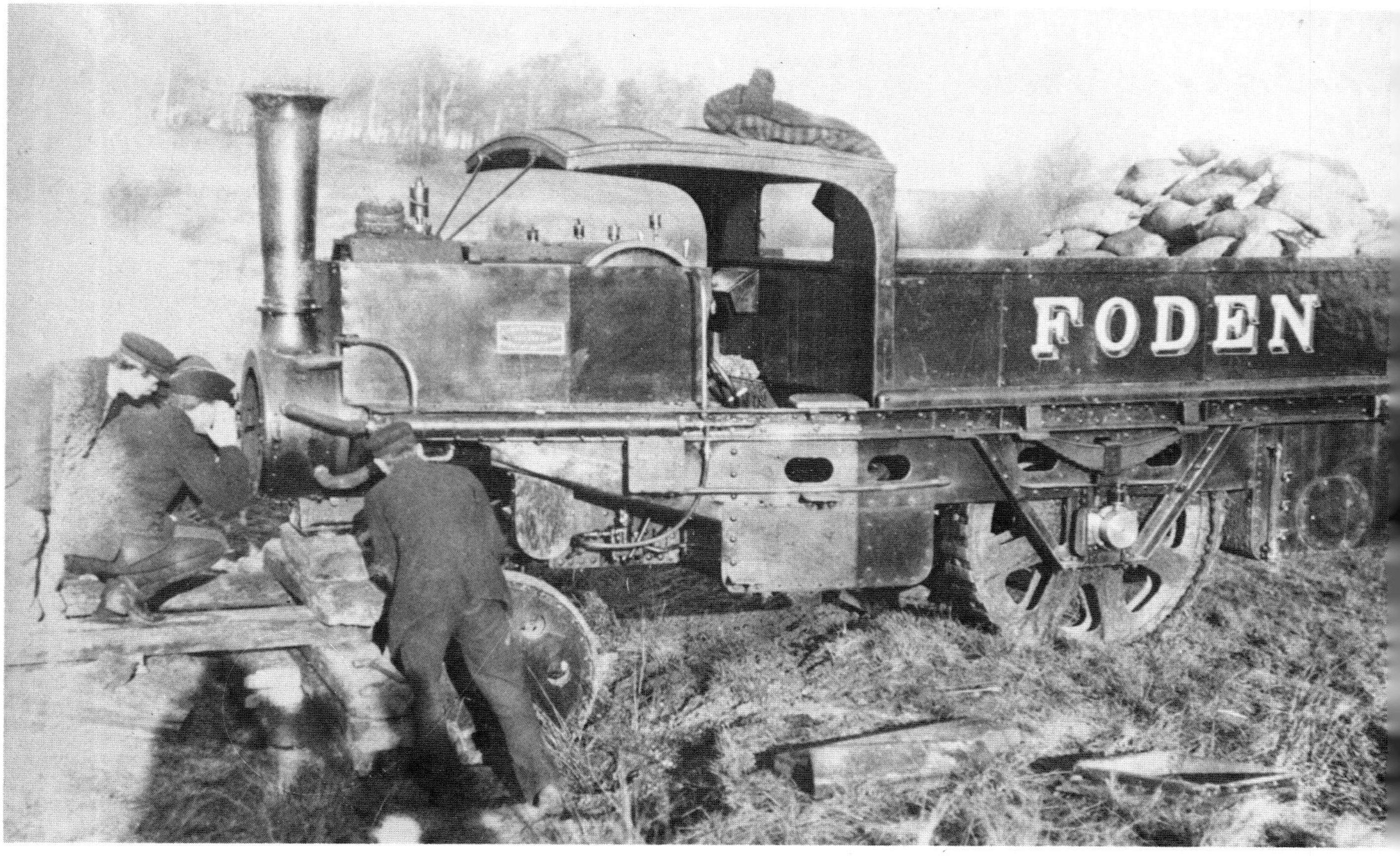
FODEN

23 Of all the French steam wagon builders Valentin Purrey of Bordeaux, the last to come to the market, was the most successful, founding a firm that continues to be in business and which built road steamers until c1930. All his vehicles had a dry backed vertical water tube boiler, but whereas earlier versions had a four cylinder tandem compound engine, similar to that which he used in his steam trams and steam railcars, later he used a twin cylinder simple expansion engine. About a dozen Purreys ran in this country, including the famous van owned by Harrods of Knightsbridge.

24 At the 1887 Royal Show at Newcastle, Edwin Foden showed an 8 nhp compound traction engine working at the then unprecedented boiler pressure of 300 psi. So well did it perform and so well was it made that Foden became noted for high grade, but dear, traction engines. At the close of the last century he began to experiment with steam wagons. He built and discarded an undertype and, in 1900, made an overtype using a wet bottomed launch type boiler which was followed by a similar design using a conventional locomotive boiler as shown. Entered in the War Office trials of 1901, this was let down by the tandem chains that had to be used between the gear shaft and the rear axle differential, no single roller chain being available of sufficient capacity - a situation soon righted by Hans Renold, whereafter Fodens began to enjoy commercial success.

25 The 1901 War Office trials exposed entries to mud, rain, fog and very rough terrain but established that makers had arrived at designs which could stand a hammering. The vehicle shown is a Thornycroft Colonial wagon with the boiler and machinery at the rear.

26 James Mann developed Parmiter's idea of the demountable transport box on a steam tractor in a number of ways. Perhaps the natural development was a demountable 'Lancashire flat' for use in the cotton and woollen trades on their respective sides of the Pennines. Mann had a spell of some eight years during which he made both undertypes and overtypes, all with locomotive boilers. The vehicle shown was made for Hattersleys, the Keighley loom makers, and the flat was detachable, though a good deal of dexterity and a certain amount of bad language were needed for recoupling the flat to the unit.

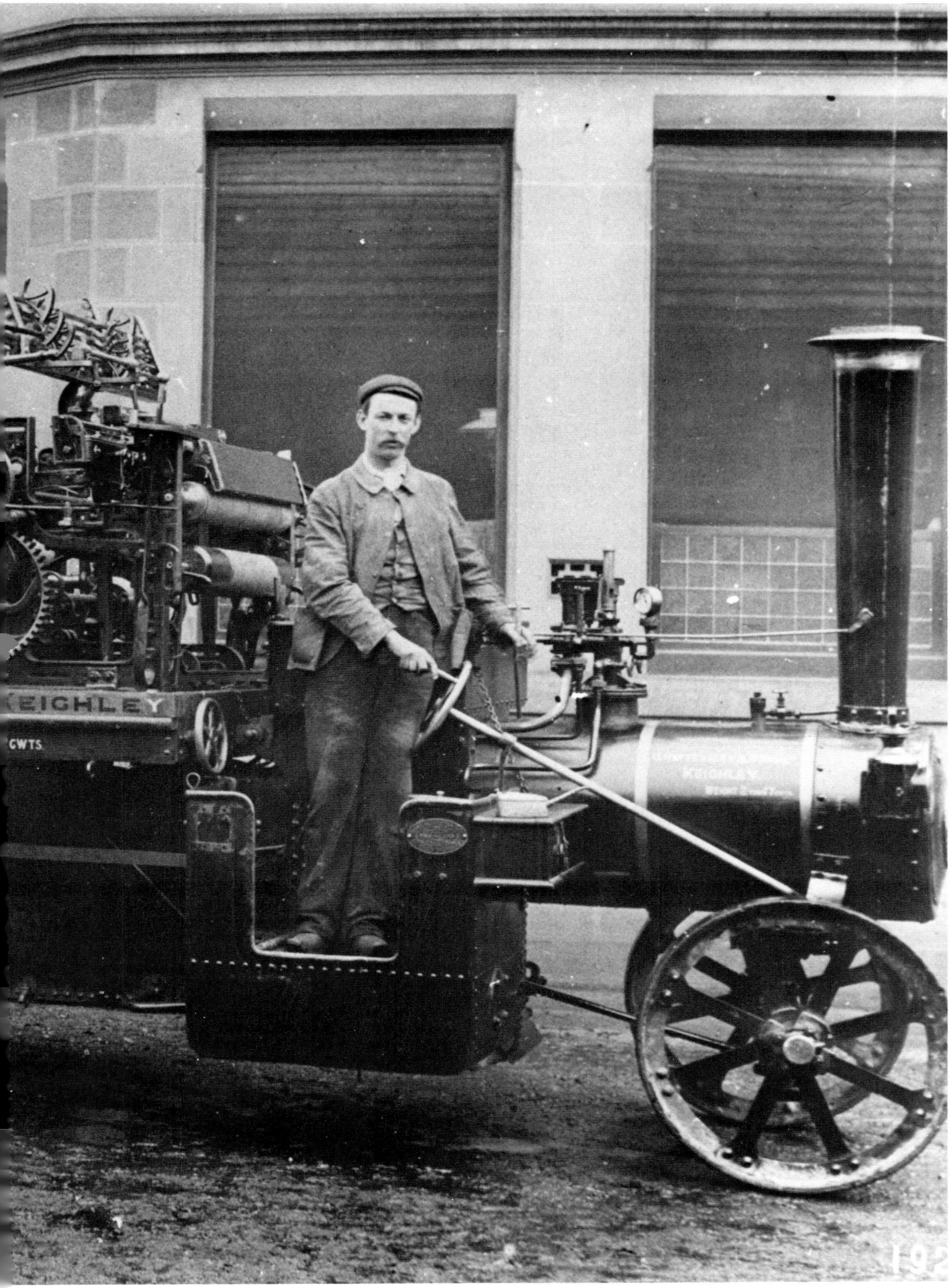
KEIGHLEY
CWTS

THE DIVERGENT STREAMS

The War Office trials of 1901 established, before a panel of judges who could not, by the wildest stretching of the imagination, be held to have any vested interest in wagons, that there were certain steam lorries available capable of sustained hard work in rough conditions. It demonstrated also that there were other steam wagons not in this category, a point not so widely heeded.

The rise in interest in mechanical traction amongst commercial users and the promise of a lucrative market led many firms and individuals to embark upon steam wagon building, often with inadequate plant or facilities or with insufficient investigation of the problems they were tackling. What ought to have been prototypes subjected to a long process of testing and improvement were launched direct upon the market with corresponding disappointment to their purchasers.

Few technical journalists of the period had any extended practical experience of the subject of wagons and, under the influence of good lunches and carefully contrived test runs, printed more or less verbatim the optimistic handouts of the makers. One looks in vain, in most of the writings on steam wagons at that time, for any real appraisal of the good and bad points of the vehicles offered or any evaluation that would assist potential buyers. Some were very sound, like Foden No 726, delivered to G Brotherwood & Sons in Tonbridge, Kent, in 1904, which remained at work for its fourth owner until 1934. Others, such as the Lomax, hardly performed one month of satisfactory work for their owners in their brief lives.

28 Foden No 726 [delivered in March 1904] in the hands of its first owners, G Brotherwood & Sons of Tonbridge, Kent. After passing through the hands of two intermediate owners it came into the ownership of the Wenlock Brewery Co Ltd in London, for whom it ran until 1934. The photograph was taken in Medway Wharf, Tonbridge, where Brotherwoods had their depot.

27 The progressive firms who might have been expected to take up steam traction were those who often had the best and most modern horse-drawn rolling stock. In France, De Dion & Bouton and F Weidnecht both made avant-train tractor units to fit under a horse-drawn lorry or bus in place of the forecarriage and just as the idea appeared to have been dropped in France, it cropped up here. The Carmont, designed by H Carmont of Kingston-Upon-Thames, appeared in 1901, to reappear as the Centre Steer in 1903, from an address in Brighton. A preview for the press, announced with much ballyhoo, was cancelled at the last moment and the project was forgotten, only to reappear as the Lomax in 1906, launched by John Goode from an office in Bishopsgate Street, London. It was a twin cylinder engine geared to a counter shaft containing the differential from which twin chains drove the front wheels. That illustrated, perhaps the only example, was sold to Groves & Whitnall Ltd, the Salford [Lancs] brewers, and is believed to have been a total failure.

29 Life was not all work even in 1900. Simpson & Bodman of Cornbrook, Manchester, gave their employees a half day off to celebrate the Queen's birthday and took them for a thirty mile trip into the Cheshire countryside in two steam wagons of their make. Apart from jamming the chimney of one wagon under the Bridgwater Canal Bridge near Dunham Park, without doing much damage, they arrived safely at the Swan at Bollington for tea and speeches. On the return trip, the copper main steam pipe of one of the wagons fractured but all eventually got home safely. Simpson & Bodman used a semi-flash boiler of their own design and, at the time of the picture, were using separate engines for each rear wheel. They were brass plate engineers and virtually everything about their wagons was sub-contracted.

30 E H Gillett's Gillett Motor Co of Hounslow, Middlesex, supplied a double decked steam bus to H J Lawson's Motor Omnibus Syndicate Ltd in 1898, but it never saw service in London though it may have run in the Thames Valley. His best known vehicle was the immaculate steam van supplied to Warings, the Oxford Street furnishers in 1903. The vertical water tube boiler was paraffin fired and had a downtake flue. The piston valve compound engine with Joy's valve gear was combined in a single oil tight [so it is said] casing with the rear axle. After six years use, Warings rebuilt the van in their motor shop but sold it in 1910.

31 Germany did not produce many steam wagons. The 1906 two ton Stoltz was made by the Hannoversche Maschinenbau Actien Gesellschaft of Hannover. The principal novelty was in the boiler design which was a twentieth century version of the Hancock boiler, not unlike six domestic steel radiators coupled side by side over a coke fire. The makers hoped to interest the German Army but the latter cannily preferred internal combustion engines. The Stoltz was made also in France under licence as the SAGE

To attempt to classify wagon types into even the broadest groups presents problems. Overtypes and undertypes, with a third category labelled 'others', is perhaps the simplest primary classification. All overtypes had locomotive boilers and were compounds showing strong evidence of descent from the traction engine. The boilers for undertypes were extraordinarily diverse. Mann built undertypes with locomotive boilers, a great many makers used various forms of fire tube vertical boiler and a lesser number used cross water tube vertical boilers. Hercules used a transverse double ended quasi-locomotive boiler in conjunction with an undertype engine but Yorkshire used it also with a vertical engine. There were also double and single drum water tube boilers with dry backs and oddities like the Beyer Peacock boiler - nominally a locomotive type but with the horizontal barrel reduced to vestigial proportions. A few makers persevered with flash steam.

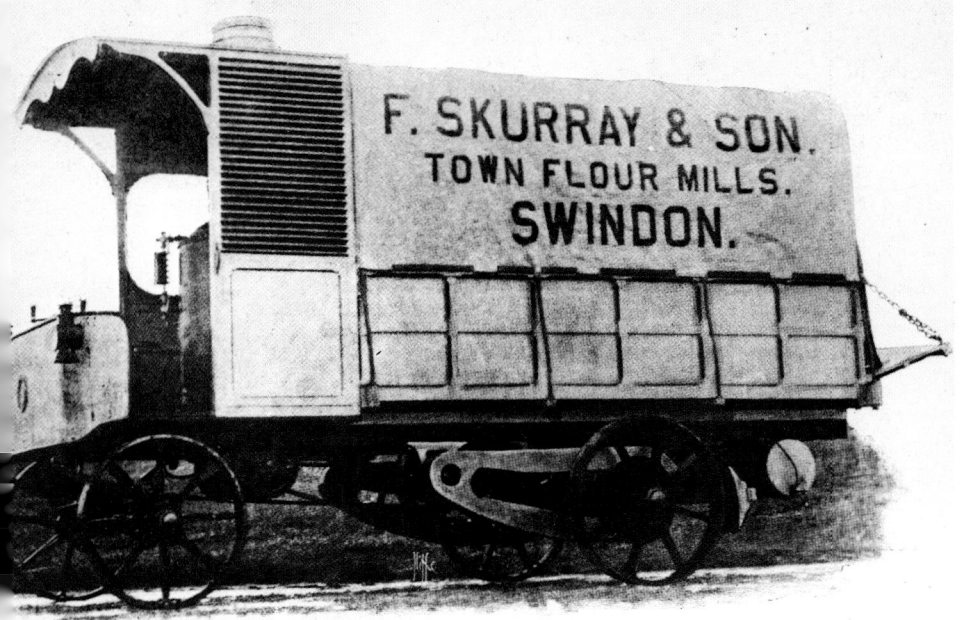

33 When the 1904 Heavy Motor Car Order raised the tare weight limit to 5 tons, many makers were ready with designs to take advantage of the easier restriction. James & Frederick Howard of Bedford had a large and old established business in ploughs and agricultural machines and a limited trade in traction and ploughing engines. Their steam wagon, first offered in the Spring of 1904, was, unfortunately, a typical fire tube vertical boiler compound undertype, a sluggish and wet runner, liable to be winded by sudden steam demands.

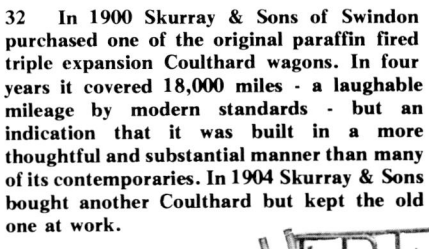

32 In 1900 Skurray & Sons of Swindon purchased one of the original paraffin fired triple expansion Coulthard wagons. In four years it covered 18,000 miles - a laughable mileage by modern standards - but an indication that it was built in a more thoughtful and substantial manner than many of its contemporaries. In 1904 Skurray & Sons bought another Coulthard but kept the old one at work.

34 Serpollet was chronically handicapped by lack of access to capital until he teamed up with the well-to-do American, Gardner. His steam trams had been in use in Paris for many years and were effective but very noisy, involving the General company in conflicts with the municipal authority. In 1905 he produced the 2½ tonner steam lorry in which he hoped to interest buyers of lighter vehicles. To his disappointment, they preferred petrol vehicles.

35 A man on a hobby horse can seldom be persuaded to dismount. After their rebuff by the General, Gardner and Serpollet joined the energetic and unpredictable Alexandre Darracq in making Darracq-Serpollet steam buses and promoting operating companies to use them - a system later used by Clarkson. In London Darracq-Serpollets were run by the Metropolitan Steam Omnibus Co, but proved underpowered on hills. Serpollet died in 1907, Darracq quarrelled with the backers, and the bus companies were put into liquidation. The Metropolitan routes went into the fold of the London General and petrol buses replaced the steamers.

36 Turners of Wolverhampton, well known for their activities in the motor car field, were also licensees of the Belgian Miesse steam cars and light commercials. The 30 cwt example shown worked in Japan. It had a three cylinder single acting engine designed on motor car principles and the dry-backed paraffin fired vertical boiler had a central pot acting as a head for horizontally coiled water tubes. Turners lost heart c1912 in the face of the continued preference of commercial steam users for the simple but reliable overtype.

37 In his seventies Colonel Francis Sheppee, a wealthy director of several nationally known companies and chairman of Exchange Telegraph, became a director and shareholder of the Power, Traction & Lighting Company in York, making, *inter alia*, Serpollet cars, from which he withdrew to set up his own company. By 1910 he was able to offer vehicles like this brewer's dray for the Tadcaster Tower Brewery Co Ltd, which were neat, fast and very well engineered. The market, however, remained aloof and no Sheppees were made after 1914.

38 Hindley's works was in the pretty village of Bourton, Dorset, where they made a variety of fixed steam engines and ship's auxiliaries. After building a few undertypes with fire tubed vertical boilers, they turned to the locomotive boiler with circular firebox, as in this wagon, for increased thermal reserve. The engine was vertically inclined and placed immediately behind the firebox which had a large 'hay-cock' steam space. Though the design looked promising, the makers did not persist with it and ceased wagon making about 1909.

39 Frank Bretherton and L C Bryan had a haulage and engineering business in Willesden Green, London. Bretherton, the engineer of the partnership, designed this wagon in 1905, using a top fired locomotive boiler, a compound undertype engine and an all-gear drive. The wagons were built by Davey Paxman & Co Ltd, Colchester, Essex. One was used by Barrett & Co, the Wood Green sweet makers, and another by Finchs of wine bar fame, but Bretherton, who was a thoroughly practical designer, concluded about 1908 that his design had no advantages over the overtype. Later he designed overtypes for Robey and Ransomes.

40 By 1907 Thornycroft, who had sold several hundred of their wagons where many other makers had been able to sell only one or two, reached the conclusion that their petrol vehicles offered better prospects for development than steamers. After 1907, Thornycroft steamers were made only by Stewart & Co [1902] Ltd of Glasgow, under licence, and sold as the Stewart Thornycroft. Later Thornycrofts used a short locomotive type boiler, as in this 1910 picture of one in use by Farnham [Surrey] Brewery, but the vertical water tube boiler was used in a massive colonial design with the machinery at the rear.

41 By 1904 Foden ceased to use the large rear wheels, placed inside the frame, which had characterised earlier wagons, and were building their classic 5 tonner - sturdy, reliable and easy to repair. Fodens were beloved of brewers and they continued to be used by London brewers until the late forties. In this picture, Driver McCafferty and his mate pose outside *The Wheatsheaf* at Loose [Maidstone]. Half an hour or so before the wagon had been in collision, in fog, with a Maidstone tram. The wagon continued its journey but it took half a day to move the tram.

42 Except for an early wagon by Komarek of Vienna, who made steam railcars, there was no building of steam wagons in Austria, and when Prince Colloredo Mannsfeld of Dobrisch Castle required a wagon for forestry use, he turned to Mann of Leeds, who sold him this Colonial 6 tonner.

43 E W Rudd, the London haulage contractor, bought steam tractors from Garretts of Leiston and in 1908/09 collaborated with them in the design of the Garrett overtype 5 tonners, the first of which [No 27720] is seen here soon after delivery. The Garrett was a good wagon, second only to the Foden, and that only by a very narrow margin. Subsequently, Rudd had a considerable fleet of Garretts which he used on the frozen meat run from the London docks to Smithfield market.

44 Taskers of Andover went through many vicissitudes and several reconstructions in their long history. Technically, the period 1905-1914 was one of their good spells, during which they produced their neat and efficient 'Little Giant' steam tractor and their overtype wagon. The wagon was not as good as the tractor. Taskers lacked the financial resources to amend the design in the light of experience. The wagons were workers and better than most of the tired compound undertypes, but did not have the capacity for sustained slogging which the Fodens and Garretts possessed.

SCOTTISH CO-OPERATIVE WHOLESA
95 MORRISON
GLASGOW
No3
V-513

W. GRAINGER & SON
HAULAGE CONTRACTORS
TUNBRIDGE WELLS
FURNITURE REMOVED TO ALL PARTS
OFFICES, 38, St JAMES PARK.
W. GRAINGER & SON, HAULAGE CONTRACTORS
01330

45 The success of Alley & McLellan as builders of ship's machinery owed much to the talents of Stephen Alley as an engineer and businessman. In 1906, having noted the failings of the many undertypes launched upon the market after 1904, he produced his own 'Sentinel' design of an effective undertype. His formula was a cross water tube boiler, proper superheat, an ample two cylinder simple engine with cam operated mushroom valves and a single speed, all allied to good workmanship and strength where it was needed. He had evolved an undertype as rugged and reliable as the overtypes.

46 Early Yorkshire steam wagons, like this van used in Walter Grainger's furniture removing and carrying business in Tunbridge Wells, were compound undertypes, made notable by their use of the Yorkshire double ended boiler with central firebox. The flue gases went outward from the centre to the separate smokeboxes and returned to a central smoke header chamber above and separate from the top of the firebox. Earlier models had double smokebox doors into which the exhaust was delivered by articulated copper pipes discharging again through a series of nozzles, each nozzle pointing up one of the smoke tubes that led to the header. Later wagons exhausted into the chimney base and seemed to steam just as well. Later Yorkshire mounted the engine vertically in the back of the cab. According to Mr Grainger's son, the old wagon worked reasonably well but was heavier on coal and water than the vertical engined Yorkshires they owned.

47 At the Berlin Motor Show in November 1906, Freibahn Gesellschaft of Seigefeld near Spandau showed a large steam wagon cum tractor, designed as the first unit of a train of two wheeled trailers, each pair of wheels driven through a prop shaft from the wagon. The boiler was a flash, or semi-flash, type oil fired on the Hydroleum system [patented in London by the Hydroleum Motor Co] under which heavy oil was injected with a jet of steam into a cone shaped casing within the firebox. All axles had roller bearings. The trailers were steerable for reversing.

The Heavy Motor Car Order 1904 added further yeast to ferment the brew still more. For a couple of years hardly a week went by that did not produce a new or revised design, mostly undertypes. The high proportion of failures or semi-failures among the undertypes gave the whole genus a tarnished image in the eyes of buyers and allowed the overtype to dominate the market. Most users understood, or imagined they understood, how the overtype should be maintained and driven but far too many undertypes required constant attention and maintenance. In 1906 Ross & Glendining Ltd, Worsted and Woollen Millers of Dunedin, New Zealand, bought a new Straker. Mr C Ovens who drove it recalled, in 1948, '..... this vehicle was the cause of many a worry as it was continually requiring attention and repair. Often work went on throughout the night in order that the wagon would be ready for the road the following day'. Thornycroft moved out of steam in 1907 and it was largely the work of Stephen Alley, with his Sentinel design, that redeemed the undertype from 1906 onwards.

By about 1910 the pattern was clear - for an overtype to succeed it had to be a Foden, or very like one, whilst for an undertype the prescription was that of Sentinel, the cross water tube boiler, two cylinder simple expansion engine and proper superheat.

During the decade the steam wagon advanced from being a machine which only bold innovators aspired to own to the status of a regular tool for the practical man. By the time registration of motor vehicles was introduced in 1904 the number of goods vehicles in use had exceeded 4000, of which about one in five was a steamer. By 1910 it had passed thirty thousand in total, but most were of not more than two tons carrying capacity and the proportion of steamers had fallen to around 5%, though these formed the backbone of heavy and long distance haulage.

48 Leyland revised their designs in 1901 when they introduced a vertical fire tube boiler for solid fuel, coupled to a compound undertype engine with Stephenson link motion reversing gear. The design was again revised in the 1905 five tonner to take advantage of the weight concession and Joy's gear substituted for Stephenson link motion. Though not notable performers on the open road, Leylands were very reliable and were very popular in short distance work with local authorities, in which they were assisted by the makers' willingness to undertake maintenance on a contract basis for a fixed annual sum.

49 The *avant-train* idea died hard. Dr A W Brightmore of Egham Hill, Surrey, first had the idea of his two wheeled steering and propelling unit about 1901. The first one was made for him by Coulthards of Preston but they declined to take up his invention as part of their stock-in-trade. In 1905 it reappeared at the show at Agricultural Hall, Islington, under the name 'Manchester' made by Turner, Atherton & Co Ltd of Denton, Manchester, but it was not heard of again.

50 The Aveling & Porter overtype, like the Garrett, first appeared in 1909. Two hundred and ninety two were made altogether - the last twelve being made, after the formation of the AGE combine, by Richard Garrett & Sons Ltd at Leiston Works. Avelings were quite acceptable performers but never caught up with Fodens or Garretts in commercial esteem, as reflected in sales. The wagon shown is No 7241, operated by Spillers & Bakers Ltd, Cardiff.

50A William Whiteley was an objectionable man, obsequious to his superiors and arrogant and sadistic to his inferiors, amongst whom he counted his staff. He met his end by being murdered by a young man claiming to be his natural son. Notwithstanding all this, he was a man of perception in selecting both his merchandise and his equipment, singling out the Thornycroft as a practical and workable vehicle from the large number of indifferent or downright useless makes with which the market was beset about 1904/05. Most of the London stores who had owned steam wagons went over to motors before 1914, preferring the greater cleanliness of the latter.

THE REGULAR WORKHORSE

**When King George V came to the throne in 1910 the national outlook had already been purged of some of its Victorian fetters by the short reign of that genial and cynical monarch Edward VII. Whilst the monarch and court did not rule the country more people than acknowledged the fact drew their standards of acceptable behaviour from the public attitudes of the King.
King Edward had self-evidently approved of travelling in luxurious cars.
His earnest son went further by multiplying the commercial vehicles in use on the Royal estates.**

It would be fanciful to conclude, however, that Royal attitudes to motors alone were responsible for the fact that in the first five years of the new reign the number of registered goods vehicles went up from thirty thousand to close on eighty-five thousand.

The five tonner rapidly became

51 Though the original St Pancras wagons [as above - by the St Pancras Ironworks Co Ltd, Holloway, London] designed in anticipation of the 1904 changes in the regulations, used a less than satisfactory fire tube boiler, the example above had a long life and its carcase lay in a Lymington [Hants] scrapyard until the fifties. The undertype compound engine had a piston valve on the high pressure side and a slide valve on the low pressure. Two countershafts were used and the differential was in the second. These wagons had reasonable success in town cartage but not sufficient to tempt their makers, who had a large business in general engineering and ironfoundry, to persist in making them.

52 When the original manufacturers of the St Pancras wagon ceased to be interested in it, about 1911, the designs and patents were taken over by the S M Car Syndicate of Hythe Road, Willesden Junction, London, who had already marketed in 1910 a shaft drive lorry using a paraffin fired semi-flash boiler under the driver's seat and a two cylinder inline vertical engine placed under an i.c. type bonnet with final drive by shaft through a worm drive back axle bought in from Dennis Brothers of Guildford. Like the Sheppee, which it superficially resembled, it was largely ignored by purchasers.

established as the standard size of steam wagon and most others were three tonners, with a few two tonners. Smaller sizes were catered for virtually exclusively by petrol motors and sales of the smaller steam lorry or van, never numerous, had ceased to be canvassed by 1910. The upper limit of 5 tons was mainly the result of weight restrictions in legislation and in local bye-laws. In France, by contrast, Purrey was building and selling ten tonners. As early as 1906 the Say sugar refineries in Paris were using sixteen ten ton Purreys. Except for Purrey, all the French builders had dropped out by 1910. De Dion defected to motors in 1904 and the last recorded Scotte, a steam ambulance and train for the French Red Cross, was made in 1909, whereafter the indomitable Maconnais returned to his original trade of hatter at Epernay, where he lived to a great age.

53 The Marquis of Londonderry had extensive coal interests in Co Durham, served by a system of railways and by Seaham Harbour where an engine works was situated to build and maintain the machinery for his undertakings. In 1903 the works designed and made for internal use a double cylinder undertype with vertical water tube boiler, unusual in having steam events regulated by a rotary valve driven from the crankshaft but this gave place to a fire tube boiler and enclosed compound engine with overhead valves. Londonderry were one of the firms who favoured an open all gear drive. Run of the mill in many respects, they were the gainers from the very high standards of workmanship of their makers. The North Eastern Railway, possibly for reasons of commercial politics, had a fleet of them. The photo was taken in The Side, Newcastle-Upon-Tyne.

54 Sidney Straker was a London consulting engineer who had had the slightly dubious honour of being technical adviser to H J Lawson. He collaborated with Edward Bayley, a London vehicle builder, to make an undertype wagon in 1899 and soon afterwards set up a works in Bristol, making compound undertypes with De Dion boilers, supplied by Abbott & Co of Newark. As noted before, these were very heavy on maintenance but would do a day's work and some 200 were sold before 1906 when he went over wholly to the overtype design which he had first launched in 1905. The overtype boiler was redesigned in 1908 with a side firehole and a sloping back to allow a more compact layout similar to a Mann but Straker had become more interested in motor vehicles and dropped steam in 1912. This is a 2 ton Straker with a De Dion boiler.

The period was dominated by the overtype makers, led by Foden, followed by Garrett, Mann, Aveling & Porter, Straker, Tasker and Wallis & Steevens. The Sentinel usurped the Leyland as the leading undertype whilst steam buses came down to one make - Clarkson. One or two of the earlier undertypes continued to be produced for a year or two and Sheppee continued to advocate flash steam. Yorkshire continued to make their celebrated double ended boiler, now allied to a vertical engine. Drive by roller chain became all but standard. Chain and bobbin steering was used for most overtypes, though Mann used a worm and segment but undertypes had Ackermann steering.

59 Loyalty to a marque was a notable characteristic of some of the wagon users in the teens of this century, many of whom could truly be termed 'characters'. H Viney & Son of Strand Road, Preston, stayed with Leyland vehicles for many years and when Leylands declined to enter a steamer in the 1927 Commercial Motor Demonstration run of steam wagons, they put in one of their own fifteen year old Leyland steamers to show what an old wagon could do.

60 Someone once remarked that being given a Sentinel to drive after one of the numerous and lackadaisical makes of fire tube boilered compound undertypes was like waking from a bad dream. Here was a wagon that was master of its work, designed with some thought for the man who would drive it. Though only a single speeder, the engine, with 5ins bore and 6¾ins stroke in its twin cylinders, was powerful enough to take a reasonable size driving sprocket and still give flexibility, a marked improvement on many of the two or three speed types on which, as in the Jesse Ellis, for instance, the driver or mate had to crawl underneath awkwardly and, in some situations riskily, to change gear.

61 Savages Ltd of King's Lynn [Norfolk] made their name [as successors to Frederick Savage] as patentees and makers of fairground machines, but they did much else besides. Another of their activities was the making of about thirty five steam wagons. The wagon shown [built in 1904] was fitted with a Musker type water tube boiler [of which they bought the patent rights in 1903] but they would also supply wagons with locomotive boilers and vertical cross water tube boilers, though all had compound undertype engines.

60A The hill climbing qualities of the Glasgow built Standard Sentinel quickly made it a favourite in the Pennine industrial area - the Lancashire cotton and Yorkshire woollen millers and their subsidiary trades becoming enthusiastic users. The wagon in the picture was used by the Tenterhouse Bleaching & Dying Co Ltd of Norden.

62 Another Norfolk-built steam wagon was the Burrell, built at Thetford by the old established builders of traction engines and road locomotives, Charles Burrell & Sons. They built [and discarded] an experimental undertype about 1900 and did not return to wagon building until 1911 when their first compound overtype five tonner appeared, unusual for an overtype in having a countershaft differential and enclosed twin chain final drive. Another refinement was a differential lock. Twenty two of these wagons were made up to the end of their production in 1922. The firm also offered [from 1912] a conventional single chain overtype with rear-axle differential.

62A The Burrell advertisement shows their second experimental overtype [No 3289] which was subsequently delivered in July 1911 to Paramoor Ltd of Margate, in the Isle of Thanet.

BURRELL'S STEAM WAGON.

Catalogues and full particulars post free on application.

CHARLES BURRELL & SONS, Ltd., Thetford, Eng.

Telegrams: "Burrell, Thetford."   Telephone No. 6.

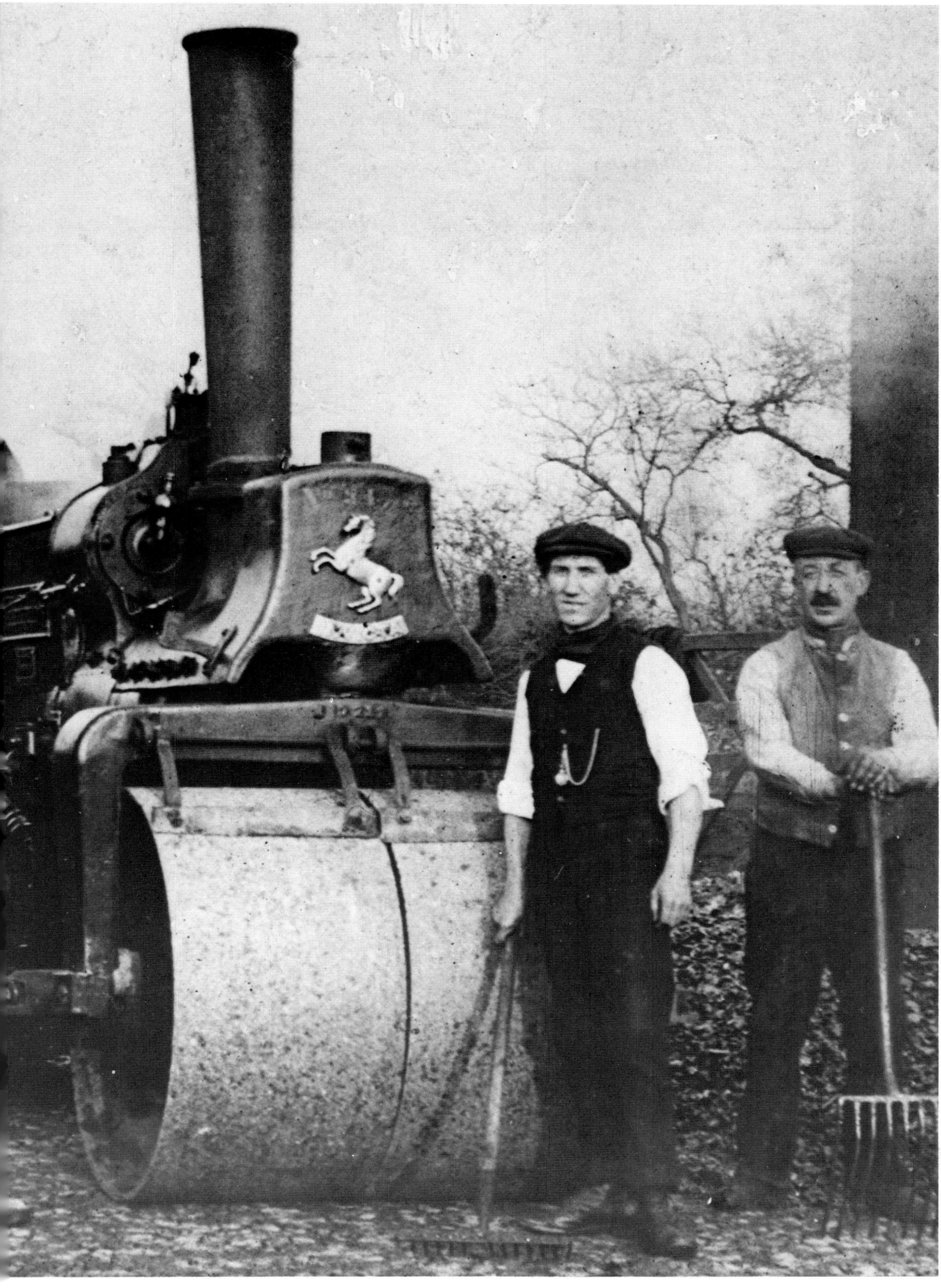

67   After the dust prevention trials at Staines [Middlesex] in 1907 under the auspices of the Road Improvement Association tar spraying was established as the soundest dust preventive treatment. Much of the tar applied annually was put on by hand-cans and brooms but mechanical sprayers were also used. Aitken's pneumatic sprayers did well at Staines and were used by the Taroads Syndicate Ltd [afterwards Taroads Ltd] who had the machines built by Manns, a connection which endured until Manns went out of business in 1928.

68   Brewers' wagons were usually smartly kept, even when old, and Foden No 9372 [new 1919] is no exception. The cast steel wheels with rubber tyres, though dear, relieved the maintenance engineer of the anxiety of keeping wooden wheels tight.

69   Clayton & Shuttleworth Ltd made their name and fame with portable engines and threshing machines but also built traction engines and steam tractors. After experiments in 1902 with a compound undertype they stayed out of the 1903/04 wagon rush and only in 1912 launched an overtype, broadly following the conventions set by Foden but using a flat topped Belpaire firebox. Most had chain and bobbin steering but a few had Ackermann. The Clayton was a tolerable performer - outshone by the Foden on road speed and hence on longer hauls and the hardest work, but a fair performer in town cartage and municipal and county council service. Many Claytons were sold in the 1914-18 war when there was a great demand for wagons and a grave shortage of Fodens.

69A It is said that Stephen Alley, apostle of the undertype, saw a very inadequate performance by an overtype and exclaimed that if he couldn't build a better overtype than that he would eat his hat [or words to that effect]. The result was the 1911 Sentinel overtype, a pioneer in having a pistol boiler - a circular firebox in a locomotive boiler - but otherwise a conventional overtype well executed. About sixteen were made and sold.

70 The 1926 design of Clayton overtype [seen here at the 1927 Royal Agricultural Show] had many advanced features - pistol boiler, internal expanding brakes on the rear axle, with the option of front wheel brakes as well, roller bearings on all wheels and the king pins, with CAV electric lighting as standard equipment - but they were tame animals compared with the Fodens which appeared more or less simultaneously. Harold Darby summed it up neatly when he said they were '..... all right for local councils and such but no use to anyone who had to cover the miles'.

The "NATIONAL" COKE-FIRED STEAM LORRY.

Write for Particulars to the
NATIONAL STEAM CAR CO., Ltd.,
16, St. Helen's Place, Bishopsgate, LONDON, E.C.

Telephone: Avenue 2301. Works: CHELMSFORD. Telegrams: "Natcarcom, Led, London."

71A Clarkson's understanding of pyschology must have been faulty. He made vehicles which, when working well, were largely automatic in operation. This made life easier for the driver at the expense of cost, both in maintenance and capital outlay, to the owner who, in general haulage, had not much interest in spending money on making life easier for the men he employed. Bowing to this reaction, Clarkson designed his thimble tube vertical boiler - two concentric cylinders forming a water jacket round the fire with short closed tubes like large thimbles protruding from the inner cylinder into the fire. These steamed well, were easy to clean and earned him an immense reputation, though more for stationary than vehicle use. Clarkson boilers were used to revitalise a number of sluggish Leylands but still failed to win the market round to Clarkson commercials.

71 Thomas Clarkson was an engineer of considerable stature. After his early wagon building with compound undertype engines and vertical firetube boilers, he went on to build chassis using his own design of semi-flash boiler, liquid firing and largely automated controls, in conjunction with 4ins x 4ins duplex cylinders, Joy's valve gear and chain drive. This combination was all very well in a private car or, to some extent, a bus, but of little interest to the average haulier. Most of the chassis sold, therefore, went under buses but disappointed even here by the sales. Clarkson decided to found an operating company, the National Steam Car Co, to run his buses in Essex, London and Wiltshire. About 200 buses of this type were used by the National. This would have been an impressive sales record, for the time, had it been achieved in an open rather than a captive market but London General, who had no ties to Clarkson, tried a dozen which they ran from Poplar depôt and would have no more. Clarkson sold out his interest in the National and steam buses ceased to be used in 1919.

72 The 1906 design of Sentinel evolved into what became known as the Standard model, which continued until superseded by the Super in 1923. Standards, though slower than the Super, were great sloggers and found their way into a diversity of trades.

73 A Mann steam cart, with Mann's own single eccentric valve gear and closely enclosed engine and gearing. The band on the rear hub is a differential lock. Steam carts were popular with brickmakers, road stone and aggregate contractors and other traders who had to cope with rough terrain.

74 The 1910 Mann three ton wagon on solid rubbers was a compact and handy tool, unusual in its side mounted footplate and side fired boiler and also in its worm and quadrant steering, more positive than the chain and bobbin steering of most overtypes. Ashbys were related by marriage to the Wallis family and also owned Wallis & Steevens wagons which were plodders by comparison with the Mann.

75 In 1911 Garretts introduced a three ton model, on which rubber tyres were standard. This wagon is No 32800 of 1915. The three ton steamers of this period found themselves in competition with an increasingly reliable range of i.c. lorries of comparable capacity and enjoyed a brief heyday of some two or three years.

76 In 1912 the details of design of the Garrett 5 tonner were revised in line with the 3 tonner. The wagons, on rubbers, were popular in Lancashire through the exertions of their agent, F E Holden of Manchester. This is No 32704 of 1915.

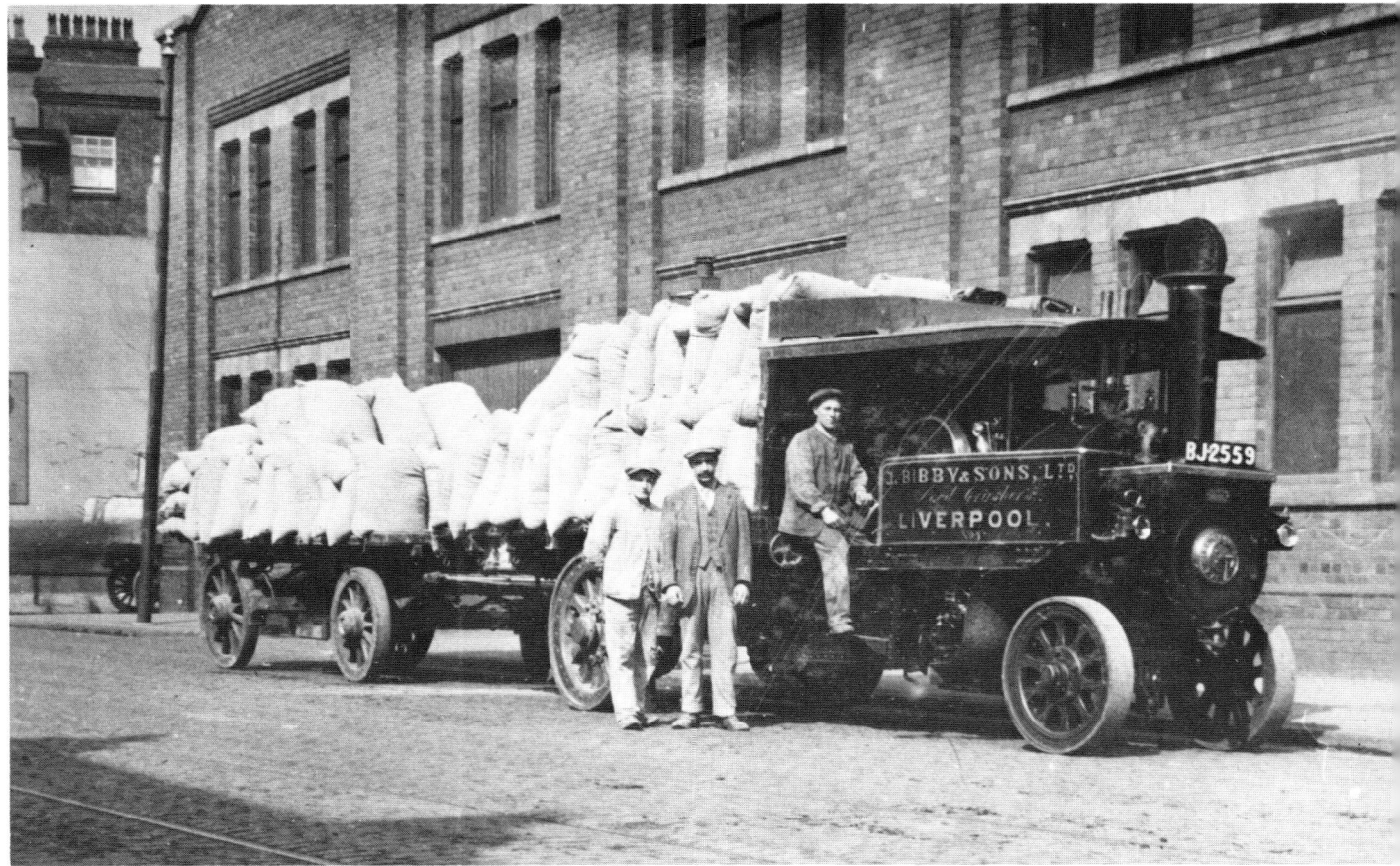

77 Aveling & Porter were heavily committed to steam rollers and were increasingly in competition with other makers. Despite this, they found time to sell a fair number of steam tractors and wagons. Aveling wagons were conventional in all except the firebox which was of the flat-topped, directly stayed, Belpaire type.

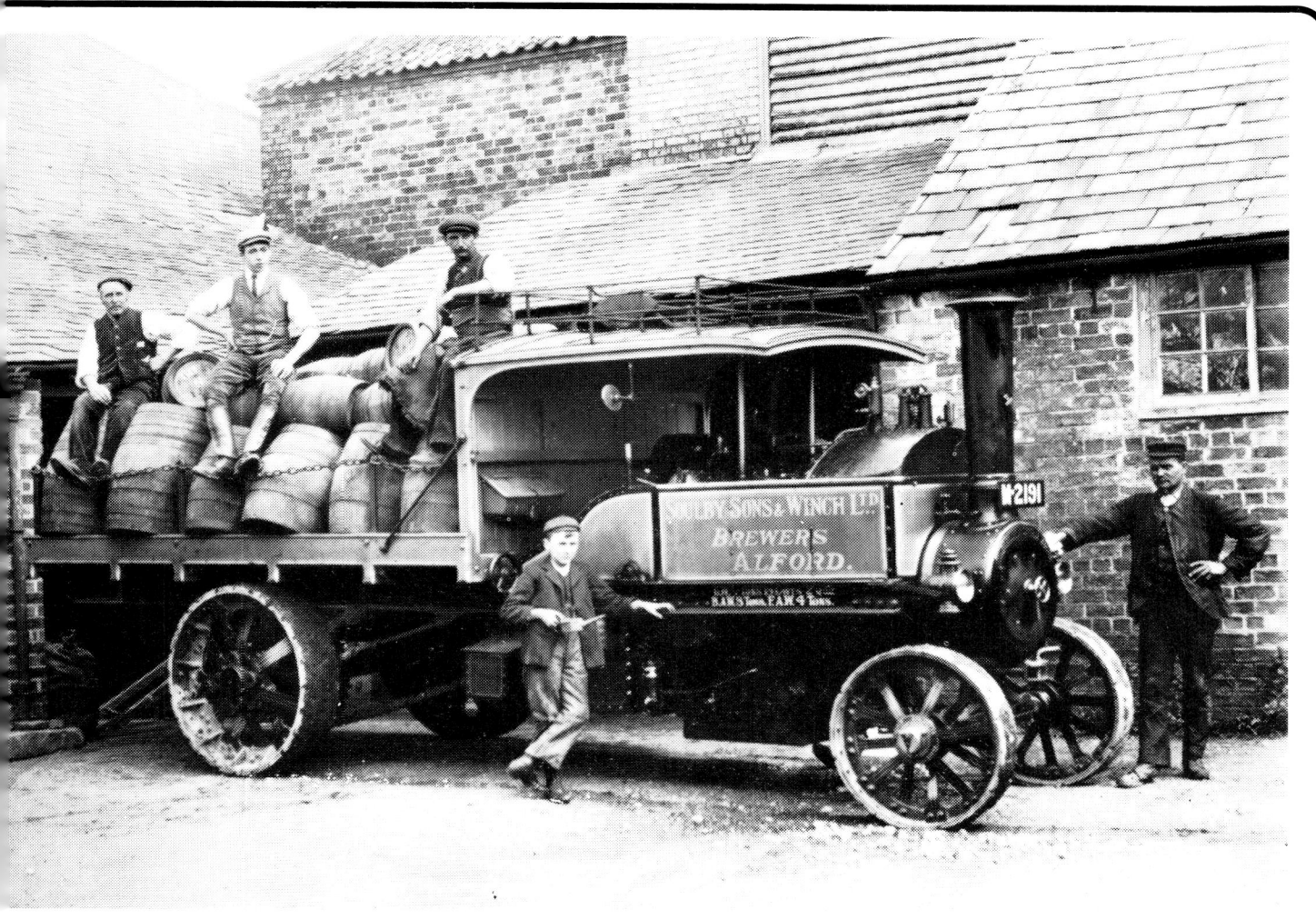

78 Fodens sold more overtypes than all the other makers put together and the 5 tonner formed the bulk of their sales. This is No 1840 owned by Soulby, Sons & Winch Ltd of Alford, Lincs.

79 In 1914 Robeys of Lincoln, rather late in the day, launched a conventional overtype design, not much different from the Foden. Of these, 15 were made but Robeys had a huge trade in fixed engines which overshadowed the wagons and it was not until after the 1914/18 War that they tackled the wagon market with an entirely new design.

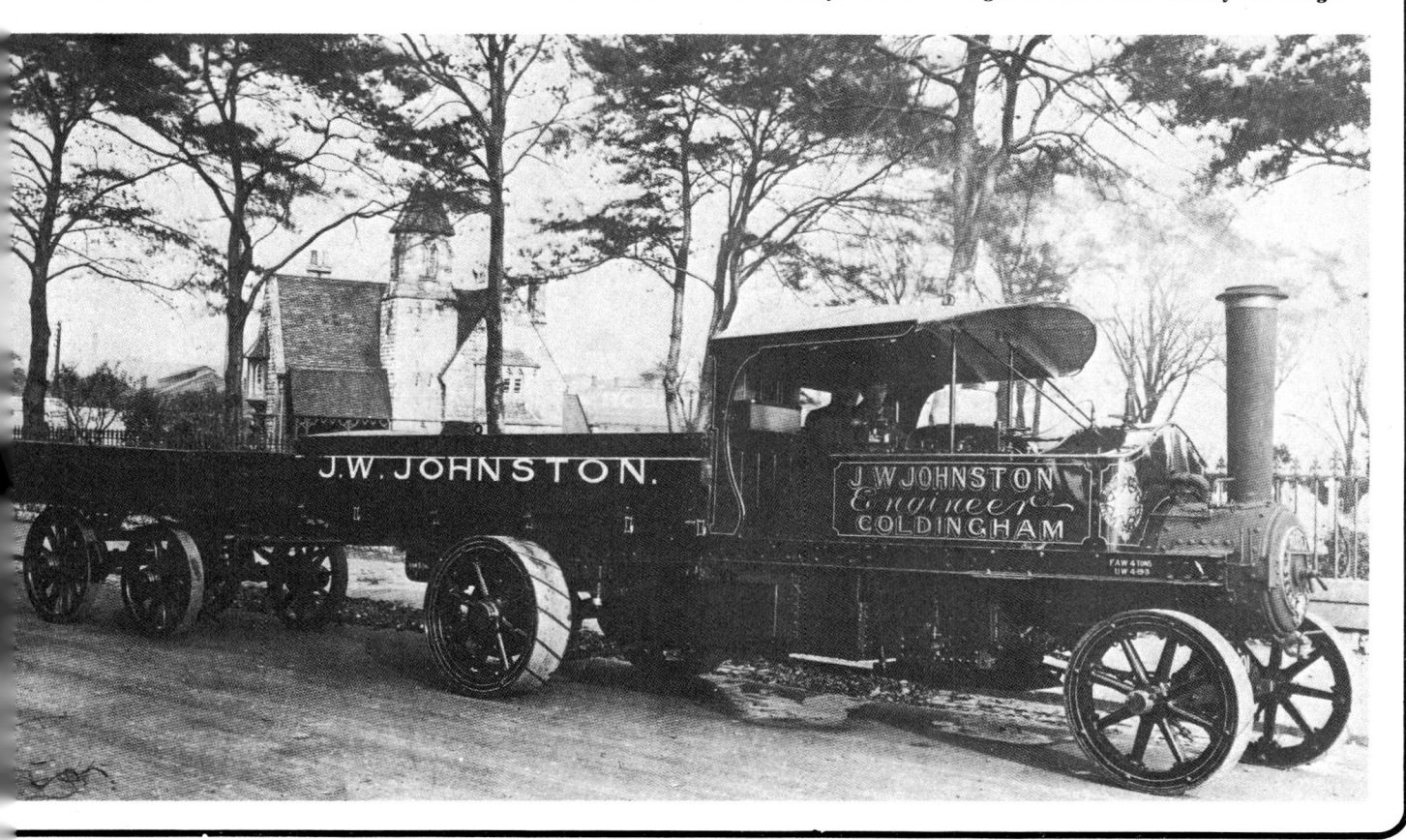

80 Despite advances in wagon design, many of the more reliable early wagons had long lives. This line-up of Mann undertype, plus trailers, used by Cardiff Corporation, all dating from the 1900's, lasted into the 1920's. The undertype Mann was a fair performer, particularly in local service, but much heavier on fuel and water than their overtypes.

81 In 1917 Garretts who had, until then, been heavily engaged in making shell cases, were given a contract for military steam wagons, many of which were delivered on their own wheels to the assembly point at Kempton Park.

82 War has always been a lousy business and until DDT appeared, battle troops soon became verminous. The drums on these Fodens are for steaming clothing to eliminate vermin and their eggs. Welcome though the respite was, it was short, for men returning to the line became reinfested from their surroundings.

83 & 83A Sentinels were the only undertypes to see military service overseas. Road building [fig 83] became very important with each advance as the road recaptured were almost totally broken up by trench building and the fury of high explosives, both shells and mines. Even in a temporarily static situation, the making good of damage by shell bursts and by sheer weight of traffic kept the road builders busy. In the lower picture [fig 83a], driver Allan Downs [arms crossed] poses in front of his Sentinel in camp at Didcot, Berks.

Local government began to use steam wagons but central government departments remained largely aloof. Brewers and millers, who had pioneered the use of steam wagons, continued to use them widely, followed by furniture removers and a large number of hauliers and general users. Retail trade used steam, ranging from Harrods' elegantly turned out Purrey, of which no good picture is available, to Tom Everton's three ton Garrett, used as a hawker's cart, from which he sold paraffin and household requisites around Droitwich.

The traffic carried by road transport during the 1914-18 war forced even the sceptics to take it seriously. Because of this increased traffic and the demands of the forces for horses, steam wagons were in great demand so that all kinds of cripples and relics changed hands at high prices. The shortage of wagons tempted Edward Atkinson, previously only a repairer, to embark upon actual manufacture. The boom hardly outlived the war. Sales of surplus army lorries at knock-out prices hampered the market in new lorries, steam or petrol, and the improvements in petrol reliability knocked out steam wagons of less than four tons capacity.

The greater road traffic density led the police forces of the country, led by the Metropolitan Police - closest to the Home Office both physically and organisationally - to press for a better field of vision for the driver of overtypes and new Construction and Use requirements in the early twenties led to radical changes in the overtype market.

84 Besides wagons owned directly by the Government, large numbers in the possession of civilian contractors were engaged on war work. The wagon above is a Burrell 5 tonner with double chain drive and three speeds - most wagons were two-speeders - built in 1915 for Robertson Bros of Woking, who were government contractors in and about the large camps in North West Surrey.

85   A line of twelve Foden 5 tonners on the way to the embarkation point for the Western Front in the summer of 1917.

85A   Leyland steamers - unlike their i.c. lorries - never saw front line service but did their share of war work at home in the hands of contractors. Latter-day Leylands were respected as soundly engineered and well constructed wagons, even if not so well provided in the steam generating department as the Sentinels.

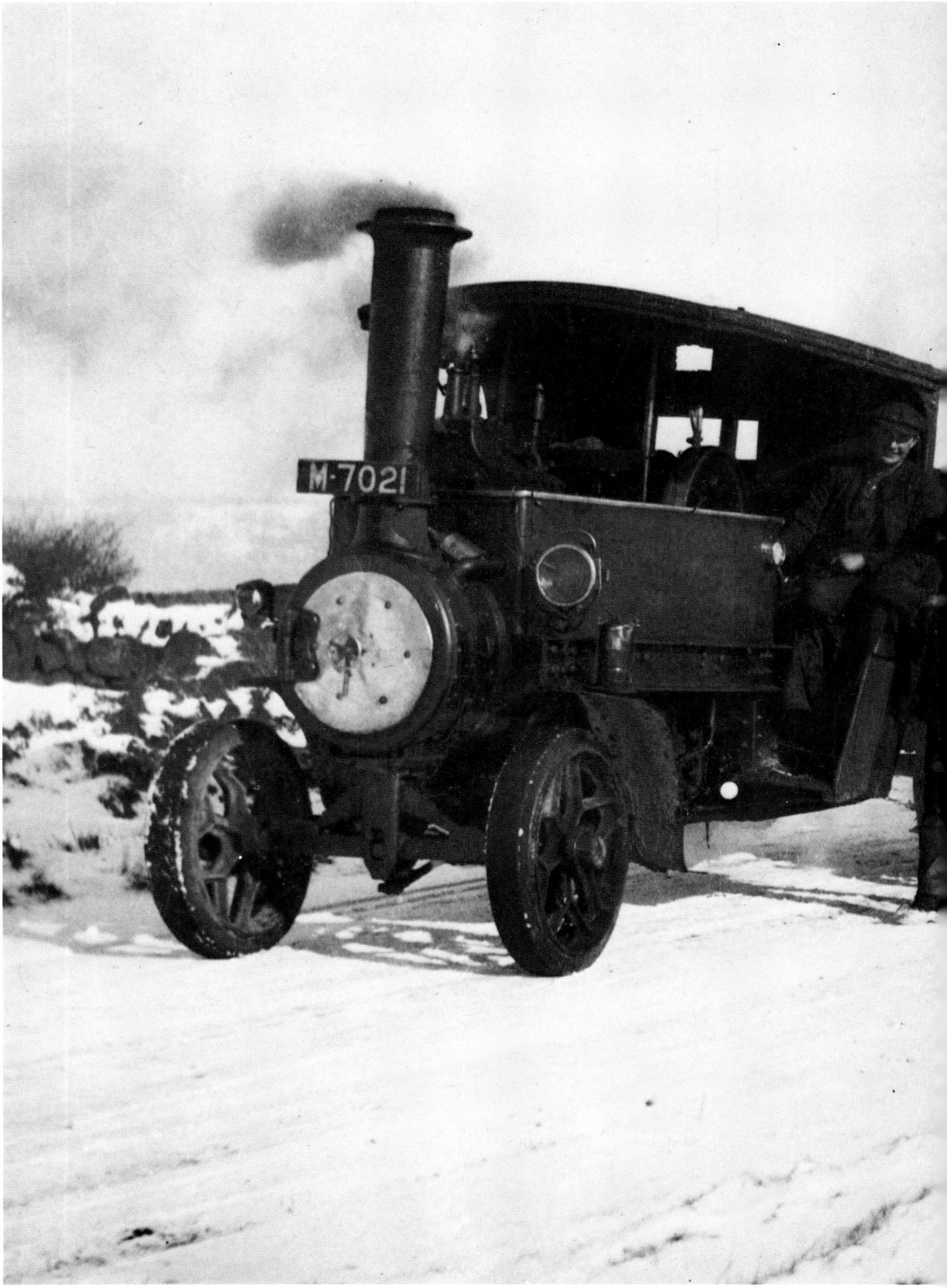
M·7021

86 The many military wagons and lorries which survived the war were released onto the civilian market in 1919. The resultant glut gave the new vehicle trade a very bad time. War Department Fodens and Garretts were snapped up by local authorities and one is shown here [*right*] working for Northumberland County Council.

The small buffers were intended to allow any wagon which became stuck whilst in convoy to be pushed by the wagon following, an idea that seldom worked in practice. The left-hand wagon belonged to H F Smith & Son of Hexham.

87 In 1916/17 the police forces let it be known that they were concerned about the field of vision from the driving position of overtype wagons and that, after the emergency, they would seek amendments to the construction and use regulations to improve the situation. Garretts produced the experimental Suffolk Punch wagon of 1917 to meet this challenge, but it was not pursued.

88 The first new design of post-war wagon was the Robey, in which Frank Bretherton had a considerable hand. A nickel steel pressed chassis saved weight. Other features were the pistol boiler with circular stayless firebox, ball bearings to the crankshaft, a high driving position which improved the field of vision, and Ackermann steering.

89 The year 1919 saw the first of sixty overtypes built by Wm Foster & Co Ltd of Lincoln, who had become justly famous for their wartime part in developing tanks. Fosters, like Aveling and Claytons, had a Belpaire firebox. The internal expanding brakes, when used in reverse, could give an ankle breaking kickback on the pedal, whilst the gear-driven pump and optional third speed were very noisy. They were, however, very well made and durable and the last overtype wagon made, in 1934, was a Foster.

The Atkinson STEAM WAGGON

6-Tonner with Trailer, 10-Ton loads all the Time
(50th REPEAT ORDER.)

Where a Great SOCIETY like this Leads YOU may SAFELY FOLLOW.

The "ATKINSONS" are characterized by:—

GREATEST POWER and CARRYING CAPACITY, ECONOMY IN FUEL WATER and OIL, RELIABILITY, & LOW UPKEEP COSTS.

PLEASE VISIT OUR STAND **No. 102** AT **OLYMPIA** October 15-23, AND SEE, IN DETAIL, ALL OUR LATEST INVENTIONS, INCLUDING OUR NEW **UNIFLOW** Engine.

:: :: Altogether an Intensely Interesting and Important EXHIBIT. :: ::

ATKINSON & CO., Frenchwood Works, PRESTON, LANCS.

Telephone—Preston 803.   Telegrams—"Waggons, Preston."

90   Steam wagons were stickers. Mann [No 1120 of 1914] in the hands of its second owner, D Pinn of Farnham, picks its way through a flooded road. After Pinn had finished with it, it went to an estate developer at Weymouth, later lay derelict for thirty years, and finally was rebuilt in the preservation era.

89A   Edward Atkinson of Atkinson & Co of Preston earned a reputation in maintaining the machines and engines of the Lancashire cotton belt and undertook the repair of steam wagons as a natural extension of his business. In 1916, spurred by the wartime shortage of wagons, he built a cross water tube boilered double cylinder single speed undertype, using hardened steel balls as valves, worked by cams, but in 1918 came out with his Uniflow engine, in which steam was admitted by cam operated inlet valves but exhausted through fixed ports in the cylinder wall. He made rather more than 300 but financial embarrassments and the small scale of production made him unable to keep up with Sentinel in development and marketing. He had a number of faithful adherents, including Bibbys in Liverpool and Glossops [the tar-sprayers] in Hipperholme, who owned over thirty wagons.

BETWEEN THE WARS 1920-39

Overtype wagon makers were faced, after the war, with the option of updating designs or dropping wagon building. Three firms who were not doing well out of wagons - Aveling & Porter, Tasker and Wallis & Steevens - chose the latter course. Robeys were the first, in 1919, with a new design, followed in 1920 by Ransomes, Sims & Jefferies, by Burrell [in 1924], Foden [in 1925] and Claytons and Garretts [both in 1926]. All used Ackermann steering and a much higher driving position. Sentinel substituted the Super [still single speed] for the Standard in 1923 and brought out the two speed DG model in 1927. Undertypes claimed a greater share of the market in the twenties, being built also by Garrett [who began to sell undertypes in 1922], Clayton & Shuttleworth [1924], Atkinson [who began in 1916], Mann [1924] and Foden [1926].

91 Coal merchants were a natural target for the steam wagon salesman and Robeys' man had a considerable success when he sold this line up to the London firm of Charringtons. Robeys were first in the field with a new overtype, launching this type in 1919, giving a much improved view of the road ahead.

92 Not only coal merchants but oil suppliers were users of wagons. Anglo-American bought Foden No 9746 [centre] and 9768 [left] in 1920 and kept them until 1930. XF 9883 [right] was an ex-WD wagon [No 8320]. Nos 8320 and 9768 were bought in 1930 by Davies Bros of Barmouth for use as tar-tankers. The wagons are lined up for the judges in a Commercial Motor Users Association parade.

In the twenties there was also a revival of interest in articulated vehicles and, in the mid-twenties, rigid six wheelers put in an appearance. Petrol was taxed from 1928 but coal was not, a temporary advantage set at nought by the Salter report, implemented from January 1934, which taxed goods vehicles on unladen weight. Fodens, weakened by family division and unsatisfactory undertypes, produced their lighter shaft drive 'Speed' model on pneumatics in 1930. Initially under-boilered it was soon equipped with a pistol shaped cross water tube boiler which gave ample steam but was regarded with suspicion by insurers. Sentinel, with better luck, experimented with a four cylinder single acting shaft drive undertype and came up with their S type, which would have won sales had there been any considerable sales potential for steam wagons.

In the early and mid-thirties, Price in New Zealand, Sentinel in England and Henschel in Germany, built advanced steam wagons on Doble's principles. The Sentinel-Dobles were used only by their builders but several Henschels were run by commercial users in Germany. Excessive cost and complication, however, placed them at a disadvantage against the ever advancing oil engine. There were sporadic outbreaks of interest in steam buses, such as the Baker in America, but all came to nothing.

93 For six years after the war Fodens continued to rely upon their 5 tonner as their staple product, but in 1925 brought out the C type 6 tonner which may be looked upon as the definitive overtype. Apart from the limitations of vision from the driving position, the overtype had much to commend it as a maid of all work in the twenties. Milehams, who owned Foden No 11774 [*above*] were haulage contracts in and about the London dockland area.

The outbreak of war and the consequent prospect of liquid fuel shortage led to surviving steamers remaining in service. Several were rescued from retirement and put back into use, such as the small fleet of Fodens run by Ind, Coope & Allsopp from their Romford (Essex) brewery.

To avoid the higher taxation on lorries, a number of sound steam wagons were cut down to tractors for use on short distance haulage and by showmen. In the twenties a number of showmen used wagons as such, but mostly went over to lorries or buses in the thirties.

94 Garretts built few overtypes after they had produced their piston valved undertype in 1922 but before the latter appeared they sold 58 wagons in 1921 and 1922. These are [*right to left*] works nos 34440, 34441 and 34443, built for Norfolk County Council and delivered on 12th February 1924.

95 Charles Burrell & Sons Ltd of Thetford had, like Garretts and Aveling & Porter, become members of the ill-fated AGE combine in 1919 by which undertype development was allocated to Garretts. Many people expected the demand for overtypes to wither away but as it became clear that a limited demand seemed likely to continue for several years, a new Burrell design was offered and appeared on the market just ahead of the Foden C type. Burrells had a great name and many loyal customers but the new wagon was a disappointment principally from its propensity for breaking axles. As William Rasbery of Gayton [Norfolk] remarked of his, 'It would break 'em nearly as quickly as you could put 'em in'.

96 When Garretts embarked on their 1922 undertype they were attempting a wagon as reliable as the Standard Sentinel but rather faster. Their first design of boiler was so nearly an infringement of the Sentinel boiler as to raise eyebrows at Shrewsbury and, in consequence, a quite fresh design of cross water tube boiler was used in production models, such as No 34358 [1923] for Premier Fish Meal. All were two-speed as against the single speed of the Sentinel.

99 A six ton Sentinel [No A2851], delivered in 1920 to C C Fyson of Bury St Edmunds and later owned by Kneller & Chandler of Bishopsgate, London. Even after the appearance of the later Super, DG and S4 wagons

many users stuck loyally to Standards, and Knellers converted their wagon to a rigid six. In the picture it had suffered the indignity of having its coal bunker catch fire outside a railway station [in Euston Road].

97 & 98 As sales of their basic overtype declined, even with a number of sound improvements, Clayton & Shuttleworth renewed the battle by turning, in 1920, to a vertical cross water tube boilered undertype with duplex high pressure cylinders and single eccentric valve gear but only a single speed. Working at 230 psi and having a superheater it ought to have been [but was not] a good worker and never won enough customers away from Sentinel to achieve substantial sales. *Fig 98* shows their experimental artic [works no UW 2037] and *Fig 99* a six ton tipper [UW 2026] sold to Scotswood Brick Co [near Newcastle-Upon-Tyne] in 1926. Only three, owned by Wingham Engineering Co Ltd in Kent, had reasonably long lives, lasting until 1939.

99A The group apparently posed with an AEC Y type chassis are, in fact, Thomas Clarkson's works staff being photographed with his final design of steam vehicle which mounted his thimble boiler and a new design of V-twin compound engine in the AEC chassis, keeping the plate clutch and gearbox of the petrol engined version. This he represented as a great advance. The clutch was one of the more expensive and vulnerable components of the motor lorry and to have it voluntarily in a vehicle that did not need it was an error unworthy of him. With a front mounted condenser in the radiator position these looked very much like motor lorries save for the chimney but less than a dozen were put to work.

100 W & J Glossop Ltd of Hipperholme were faithful users of Atkinson wagons. Their first was bought new mainly because Atkinsons were one of the few established firms interested enough in building a purpose-made chassis to Glossops' ideas, re-equipping it to their specification, but later many second-hand wagons passed through their hands. Sentinels sold them a number of Atkinsons taken in part exchange for new Sentinels and individual units lasted in the Glossop fleet up to the fifties.

101 In 1925, in defiance of probability, Garretts [who were in AGE] produced a new design of 6 ton overtype in competition both with the Foden C type and Burrells' [who were also in AGE] new overtype. The latter was not doing well but the total overtype market was shrinking and Fodens, on quality and reputation, were taking most of what there was, so that only nine of the new Garretts were sold, most of which lasted until the end of commercial steam wagons. This is No 34886 of 1926.

102 Particularly in municipal work many old wagons lasted into the twenties, such as this rare Merryweather gulley emptier of which the boiler and engine were similar to the maker's famous Fire King steam fire engine, but 1924 was a watershed for old wagons and changes in the Construction and Use Regulations brought about their end.

103 Local authorities in the twenties still appreciated the simplicity and reliability of steam. In gulley emptiers a steam ejector was a ready means of creating the vacuum on which the working of the machine depended. This is Garrett No 34669 in use as a demonstrator. It was built in 1925.

104 In the late twenties there was a revival of the early *avant-train* idea prompted by Knox in the USA and Scammell in Britain. Atkinson, Foden, Robey and Sentinel built six-wheeled artics. The engine above, owned by Fred Darby & Sons of Sutton, Cambridgeshire, was hauling sugar beet.

105 Allchins of Northampton, as in many other things, followed Fodens' lead in articulated six wheelers. This is No 1355 or 1356, new in May 1926. All the steam firms, in the end, admitted defeat with artics and turned to the rigid six, pioneered by Garretts.

106 Garrett built the first rigid six steam wagon [No 34902] in 1926 using the same cab, boiler and piston valved engine as in the contemporary four wheelers. Though subsequently sold [in 1928] to J E Heath of Nottingham, it was not repeated, and production models had a new cab with forward driving position, the new poppet valve engine and the Garrett patent bogie.

106A Shadracks' wagon [No 35432 of 1930] was a special, having a 22ft 0ins body for coke trolleying, mainly from Beckton Gas Works. Every joint in it was put together with red lead. The design of the body was done by the late Frank Waddell, then a young draughtsman in Garrett's thresher department, and Mr Shadrack was so pleased with his work that he gave him a ten shilling gratuity [equal to about twenty per cent of Frank's weekly salary at the time].

FOR DELIVERY OF COAL IN BULK

This Garrett Six-Wheeler recently delivered to Messrs. E. & A. Shadrack, the well-known coal merchants, is fitted with 22 ft. body specially designed for delivery of coal in bulk.

There is no Wagon on the market to-day to compare in first cost with the Garrett Rigid Six-Wheel Steam Wagon, in comparison with the amount of work it is capable of performing.

Built in three distinct types, the Standard, the Three-way Tipper, and the Side-Tipper, it can be seen on the roads in all parts of the country.

Haulage contractors, quarry owners, sand and ballast merchants, engineers, millers, etc., all declare that this is the Wagon, with its large earning capacity, to bring increased turnover with consequent greater margin of profit into your business.

Write for Catalogue B.655.

Write for Catalogue B.655.

RICHARD GARRETT & SONS L^{TD}
Est. 1778.
ALDWYCH HOUSE, LONDON, W.C.2
WORKS: LEISTON, SUFFOLK

ASSOCIATED WITH AGRICULTURAL & GENERAL ENGINEERS LTD.

107   Never left far behind, Sentinel soon brought out their own patent rear bogie which was fitted to the six wheeled DG model in 1927 - the first sales being in January 1928. The wagon shown belonged to Wm Woodbridge Ltd of Westminster, who owned the Thrapston pits.

108   Despite the ground lost to the undertype makers, the Foden C remained a popular wagon through the twenties. Outside the owner's depot at Hendon, No 11198 of 1924, owned by Schweppes, is being groomed for the day's work.

109 Manns revised their well tried overtype wagon, with comparatively minor improvements, into a 6 tonner in 1921, which was thoroughly dated by the mid-twenties. In 1924 they introduced their Express undertype with cross water tube boiler, superheater, two-speed gearbox, double cylinder, double acting engine and cardan shaft-drive, a sturdy and reliable, albeit rather heavy and somewhat expensive wagon, which came to the market just as Sentinel were ready with their Super. As a result, it achieved few sales and the company went into receivership in 1926 and liquidation in 1928.

110 Many of the older wagons displaced by the improved types of the twenties found their way onto the fairground. This is Garrett No 30997 owned by William Wright of Dundee.

111 John Fowler & Co [Leeds] Ltd were the pre-eminent firm in steam ploughing as well as building road locomotives, traction engines, tractors and steam rollers. Of these only steam rollers enjoyed good sales in the early twenties so that the wagon market, which they had hitherto eschewed, became more attractive. Their wagon of 1924 had a vertical fire tube boiler, vertical V-twin compound engine, cardan shaft and worm drive with three speeds and high ground clearance. High price and the inability of the boiler to cope with peak demands denied it a large share of the private market but a gulley emptier version was successful. A fleet of these was used in Warsaw [Poland].

Yorkshire Advantages No. 4.

General Construction.

112 & 113 Yorkshires never achieved the public notice they deserved. Their classic arrangement involved the well-known double ended boiler, a vertical compound engine with modified Hackworth valve gear [unusual in wagons] and final drive by long roller chain. No 1510 [fig 112] was a three speed 7 tonner gulley emptier built in 1925 for the LCC tramways. The three tonner [fig 113] was No 706 [1915], an interesting vehicle with a platform height of only 40 inches above the road, compared with about 50 inches in many other makes of comparable size. In the shaft drive alternative of the nineteen twenties, the engine was, at first, arranged in line with the gearbox and drive shaft but, in later shaft-drives, the engine was offset to one side and the gearbox beside it.

A YORKSHIRE SIX-WHEELER

THE YORKSHIRE PATENT STEAM WAGON COMPANY
HUNSLET, LEEDS, ENGLAND.
'Phones 24275, 26652 LEEDS.
'Grams "Motor," LEEDS.

LONDON OFFICE ABBEY HOUSE, 2-8 VICTORIA ST., WESTMINSTER, S.W.1.
Telephone Number 6596 VICTORIA.

Full Specification and Price sent upon application.

THE Leeds Corporation—already operating 17 YORKSHIRES—has again deemed it advisable to rely on the famous YORKSHIRE W.G., and their latest purchase of a Six-Wheeler for their Electricity Department is **good evidence** of the merits of YORKSHIRE Vehicles which are also very satisfactorily serving numerous other Municipalities and Haulage Contractors, etc. The YORKSHIRE Six-Wheeler is the ideal Vehicle for all Electrical undertakings.

SOME of its features are:—All Steel Body. Carrying capacity 12 tons. Load distributed over three Axles, thus less weight on each Axle and less wear and tear on roads. Two 7' 6" diameter Cable Drums can be carried, and other equipment. Detachable Overhead Runway is capable of holding 5 tons. Extra Low loading (1' 6") gives low centre of gravity of load. Back door is used as a ramp. Winch at front end for loading.

115 The London sand and ballast trade was well supplied with steam wagons. Here one of the Garrett four wheelers owned by the Herts Gravel and Brick Works Ltd runs down Regent Street at the end of the 25 mile run in from the pit near Welwyn. It is works no 35105 [1927].

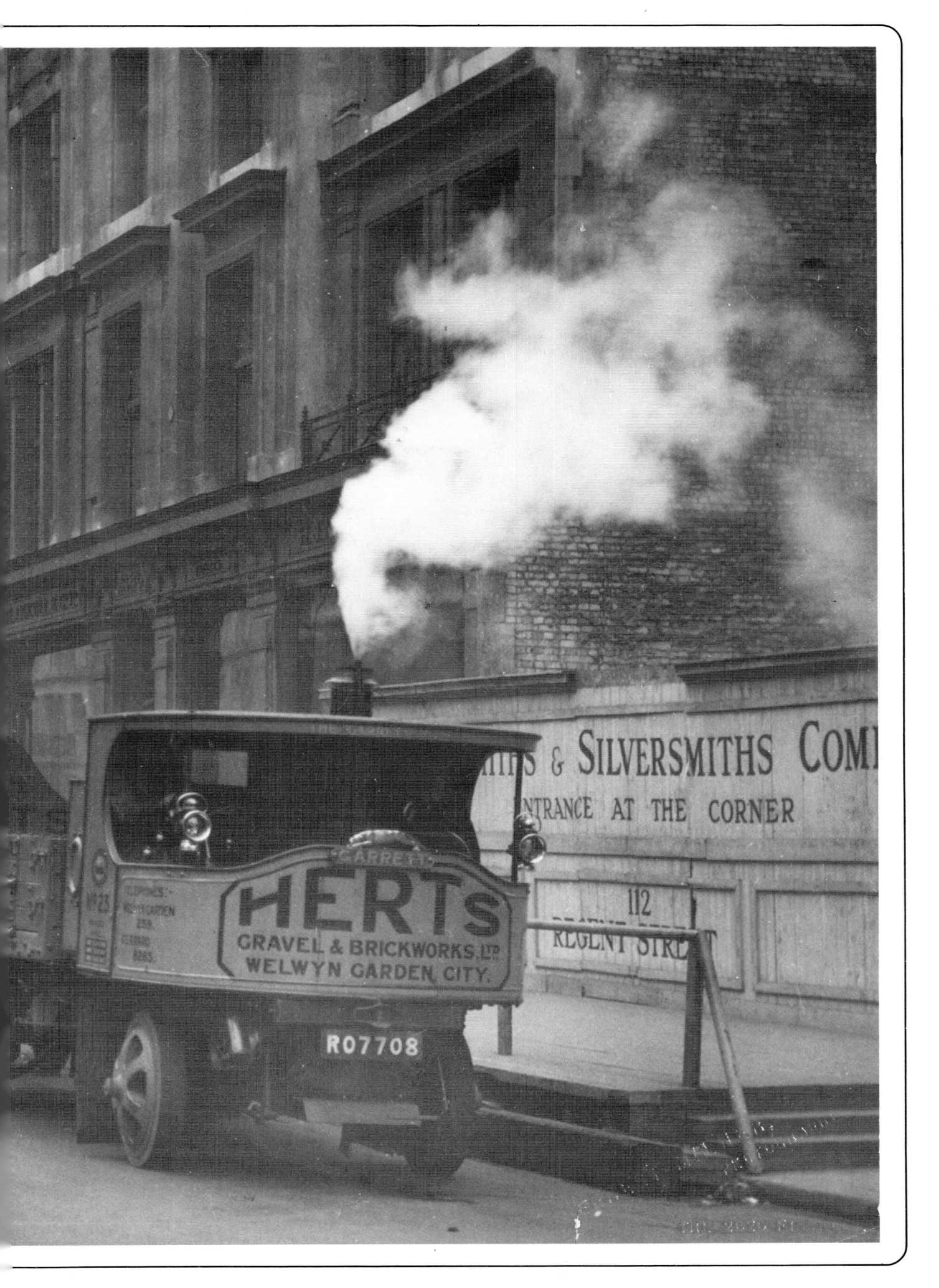

HERTs
GRAVEL & BRICKWORKS LTD
WELWYN GARDEN CITY.

RO7708

…HS & SILVERSMITHS COMP…
…NTRANCE AT THE CORNER

112
REGENT STREET

116 The load capacity of contemporary heavy duty pneumatics was a deterrent to their use on steam wagons as they brought down - at least on paper - the overall loading of the wagon. Nevertheless the economic and operational advantages were great and a number of conversions were made. The DG8 [above], owned by Tarmac Ltd, was converted by Sentinel using a new rear bogie design.

117 & 118 Both Fodens [*below left*] and Sentinel [*above*] made timber tractors based upon the boilers and engines of their wagons. Both were good machines but the Sentinel was somewhat heavier on the front wheels which made it harder to manage on rough terrain by decreasing adhesion on the driving wheels and loading the steering. The Foden was No 12670 [1929] owned by Charles Claridge of Exeter and the Sentinel was No 8369 [1930] owned by Halseys of London Colney.

119 Sentinel carried their tractor building rather further by building the Rhinoceros tractor for off-the-road service in Africa. No 6985 [1927], the prototype, was shipped to Kenya and put through its paces in very rough going but i.c. tractors gained the market and only seven were made.

120 In the thirties a number of sound old wagons were converted to tractors thereby escaping the unladen weight rule. Several firms on the London Docks to Smithfield frozen meat run, including Union Cartage, Matthews and R Cornell, converted wagons in this way. This is Matthews' Foden No 8762 [1919] which ran until 1935.

120A Wm Allchin & Co Ltd of Northampton were inheritors of a long established family business turning out about one engine a week. Before the 1914/18 war they had made a number of 5 ton overtypes following the Foden conventions - and before that a few undertypes - and continued in the twenties. Sales fell off badly about 1924 - when the C type Foden appeared - and all production ended in 1931. The total of overtypes was 256. A late type Allchin [No 1361] posed next to two Fodens [Nos 12228 and 12256] at the Royal Agricultural Show in 1926.

Foden
FAIR
WL·910
SEVENTH
REPEAT ORDER
PF·2373
E. LONGHURST & SONS. L.
THIRD
REPEAT ORDER
TESTIMONIAL

Let FODEN carry your Cargoes.

The Spacious dignity of the world and the industrious effici of the new, combine in the F Speed 6 and 12. Pneumatic ty Brakes to all wheels and ar loading space, are some of features of these Fodens which the merchant princes of to deliver their cargoes.

Write us for further particular these wagons.

THE Foden SPEED 6 & 12

Steam Wagon Makers to H.M. The King.

FODENS LIMITED, SANDBACH, CHESH.
Telegrams: "Foden, Sandbach." Telephone: 45 an

London Office:
22, Sydenham Hill Road, Sydenham, S.E
Telegrams: "Nedoluden, Forest, London."
Telephone: 2474 Sydenham.

Southern Depot: Cosham, Hants.
Telegrams: "Foden, Cosham."
Telephone: 76030 Portsmouth.

FODENS LIMITED, SANDBACH, CHESHIR

121 & 122 The Foden O [or Speed] type, the first steam wagon designed for pneumatics, was intended to equal or excel anything that the internal combustion opposition could do. The 'banana' boiler, externally like a pistol type locomotive boiler, had cross water tubes in the barrel and was placed in the wagon firebox first. A two cylinder double acting engine, two speeds, cardan shaft and worm drive, pressed steel chassis, use of aluminium wherever possible for the gearbox casing, etc, and the coach-built cab with wind-up windows made it the most advanced steamer the market had seen. Its bugbear was the suspicion with which insurers regarded it and its welded boiler. It was capable of 50 mph and of carrying gross overloads. The Bristol Co-op wagon [bottom], No 13952 [1931], was used by them until 1939 and then by Hardwicks of Ewell until 1946. The Howards & Sons Ltd wagon was No 13638. The total made was 135.

123, 124 & 125 Latterly Sentinel were having better success with their DG undertypes than Foden with the C class overtypes, but experiments with the shaft drive SD [basically like a DG] led, in 1934, to the S type in four, six and eight wheeled versions. Four cylinder single acting horizontal in-line engines with poppet valves and shaft drive were used. Pneumatics, electric light and enclosed cabs were standard and much aluminium incorporated. It was fast, powerful and versatile but less rugged than the Super or DG. The wagons shown are Nos 8821 [the first six wheeler], 8838 and 8850, the latter taken in 1961.

2
THOUSANDS OF POUNDS
WASTED DAILY
SAVE YOUR WASTE PAPER
PHILLIPS. MILLS & Co. LTD.
Government Contractors
BATTERSEA

126 Phillips Mills & Co Ltd were waste paper merchants who specialised in collecting the waste from the offices and shops of Central London. Many of their vehicles carried a headboard *'Thousands of Pounds wasted daily - Save your Waste Paper'*. The photograph of their yard at Battersea shows a line-up of steam wagons in the early twenties, mostly bought secondhand. The four tonner in the foreground [No 4454 - Registration Number M6183] was built in 1914 and Phillips Mills were its fourth owners. The second is No 3460 [1912 - Registration Number M4542].

127A & B In the thirties Abner Doble, despairing of selling steam cars based on his designs, looked to the commercial field. His ideas were taken up by Sentinel in England and Henschel und Sohn of Kassel, Germany. Henschels aimed at making the most advanced steam wagon ever and in the process secured an S type Sentinel, the then current holder of the title. In *Fig 127a* it is posed with an oil fired Henschel.

The second picture shows a production model Henschel used by Thuringia Brauerei. Operationally the wagons were great performers but were defeated by the exacting standards required to make and maintain them and by the price of fuel.

128 Doble himself worked with Sentinel's engineers to produce the coke fired Sentinel/Doble. An almost legendary performer, it failed to break the cost/habit barrier. Without sales the cost would have been high but the high cost would have inhibited sales. About 98% of road traffic was i.c. engined in 1939 and a staggering level of economy would have been necessary to break the habit for i.c. among purchasers. Sentinel looked at the figures and prospects and laid it aside though it was used as works trunk transport all through the war.

129 In the thirties many wagons gained a respite by use as tar sprayers, which were exempt, as road machines, from road fund tax, and as tar tankers. Thames Tar Products owned Sentinel No 8562 when it was photographed by John Meredith at Beddington Lane, Mitcham.

130 For many wagons the thirties meant a forlorn end in a pile of scrap. Sentinel No 8213 formerly run by Willment Bros of Twickenham languishes in the scrap pile at their depot from which it was rescued in the sixties for preservation. The body was a pioneer truck-mixer for concrete, some years ahead of its time.

CLOSING SCENES

At the end of the 1939/45 war many of the steam wagons that had been retained in service through the emergency were quickly withdrawn, many for scrap. The large Sentinel fleet of the Cement Marketing Co Ltd, the withdrawal of which had been suspended during the war, was dispersed, though some wagons saw further service with contractors.

By about 1948 the three main refuges of steam wagons were the London gas companies, the Liverpool dock traffic and the tar spraying and road making trade, though a few survivors lived on elsewhere, notably the fleet of about a dozen very old Sentinels in Brown Bayley's steelworks at Sheffield.

The London gas works moved out of steam in the early fifties, the Liverpool users gave up reluctantly, wagon by wagon, during the next five years. The bigger tar spraying firms had mostly given up steam by 1955 except for W & J Glossop of Hipperholme (Yorks) who used their last steam sprayer in the season of 1966 and kept the wagon involved (Sentinel No 8666) for historical and sentimental reasons. At the time of writing one steam wagon remains in commercial use - the Sentinel DG 4 tar sprayer operated by the North Wales firm of Lloyd Jones Bros.

131 The extensive steam fleet of the Cement Marketing Co Ltd had gradually been reduced before the war, beginning with non-standard wagons. Scrapping had begun to eat into their DGs when the outbreak of war halted the process but it took only about three years after the war to see the end of steam wagons in their service. Fleet No 81 carried Sentinel works no 8419.

132 An object lesson in how smart a sprayer could be. H V Smith & Co's No PS8 (Sentinel S4 No 9026 built in 1934 for Hants County Council) photographed by Arthur Ingram in the late 1950s. Notwithstanding the faster rate at which the S type could travel between contracts, spraying firms on the whole preferred the simplicity of the single speed Super or the robustness of the DG.

133 Yorkshire No 1511, a sister engine of the wagon No 1510 in photo 112, was a 7 ton gulley emptier used for cleaning the sumps on the tramway conduits, working from Walthamstow depôt. By the sheer accident of being still in use when war broke out, it survived until 1948, though the picture shows how its condition had deteriorated by the end of the war. Nevertheless, it was reconditioned once the war was over and spent its last years in a state of gleaming perfection, polished by a doting crew.

GAS
228
KR 9445

H.V.SMITH & Co. Ltd
TRINIDAD ASPHALT MACADAM
WORKS: TOTTENHAM & DAGENHAM DOCK
CML 734

136 Fuller, Smith & Turner of Chiswick ran steam wagons for fifty years, beginning in 1898 with a Thornycroft and ending with Foden six tonner No 13568 in 1948. This is their No 13820 at Hampton Court. In an orange-red livery it was a notable addition to the scene in the West London suburbs and delivered good beer into the bargain.

134 One of the last Garretts to run commercially was No 35454, owned by the Wandsworth and District Gas Co, unusual in having been converted to pneumatics, and still at work in 1947.

135 H V Smith & Co Ltd [now part of the Tarmac group] were post-war commercial users of steam in London. In this photo, taken in February 1955, they were working the rebuilt and re-registered Sentinel No 8454 in the Bath Road by London Airport.

137 The last Standard Sentinel was built in 1922 but examples continued to work at Brown Bayleys in Sheffield until 1966. The Standard Sentinel owned by Horridge & Cornall Ltd was converted to pneumatics and survived on local delivery work in and around Bury until the fifties.

138, 139 & 140 Liverpool, in the early fifties, was well supplied with steam wagon users - Kirkdale Haulage, Criddle & Co, the United Africa Co, [139] Bibbys, George Davies & Sons, Websters, Benjamin Sykes & Sons, [140] White Tompkins & Courage, Vernons and William Harper [138], to name some of them. The city saw the beginning of steam wagons and it saw the end.

141 Another unexpected survivor was Foden No 5788 used on internal coke transfer duties at Hilsea Gas Works, Portsmouth, until the early sixties.

142 Burning frames, being road machines, were exempt from the heavy Road Fund tax on steam wagons. This unit from the fleet of W & J Glossop is unusual in being mounted on an Atkinson chassis, one of the few Atkinsons to be converted to pneumatics.

143 Though Leyland lost interest in steam early in the twenties and built their last steamer in 1926, examples continued to be run in Liverpool, much rebuilt, until the early fifties. The wagon illustrated is F2/72/1682 [Reg No KB 8716], rebuilt as a six wheeler on pneumatics for Tate & Lyle Ltd.

144 The last throw of British steam wagon building was the 100 wagon order from Argentina to Sentinel in 1950. Here one of the wagons stands, minus a wheel, at Rio Turbio. The Sentinels had short, rough lives but served the mines until superseded in 1956 by a 750 mm gauge railway powered by Japanese-built 2-10-2 steam locomotives.

Steam lorries had a brief Indian summer in Germany when, in the desperate post-war period, Lenz & Butenuth of Berlin built a number of steam lorries using i.c. components on a more or less stop-gap basis. A more sustained experiment took place (from c1943) by Sachsenburg Brothers of Dessau-Rossban (later part of East Germany). These used a water tubed vertical boiler, hopper fired by coke, generating steam at 350 psi, which was used in a double acting single expansion engine - two cylindered in the smaller version and three in the larger. Condensers and modern bodywork completed the vehicles. The last designs were two types of tractor made in 1954, designed to haul 20 and 40 tons respectively at relatively low speeds. There was also the work of Diaz Fahrzeugbau of Berlin from 1946 until about 1950, using a horizontally opposed four cylinder engine and a solid fired watertube boiler.

145 The 1954 Sachsenburg tractors are, strictly speaking, outside our scope but their design was closely related to steam wagon practice and indicates the only likely alternative to the exacting Doble type technology of more sophisticated steam vehicles. The East Germans never published an account of their behaviour in regular work but one must assume they did not have sufficient advantages over motors to make continued production to be deemed worthwhile.

146 The Mechanical Tar Spraying Co at Reading was one of the companies that used the Fowler Woods spraying equipment actually mounted on the roller and they had Foden tanker No 13764. This picture of it at work was taken in 1961. Their steam rollers were gradually laid off, mostly as drivers died or retired, but the Foden was kept at work until the mid-sixties and was certainly the last Foden in commercial employment.

147 Henry Franklin Ltd of Biggleswade, Beds, were a firm of millers and corn merchants who used Sentinel No 8595 for trunking between their mills and the docks. It made trips to Liverpool but also visited London and, in this picture, is passing under the bridge that carried the railway siding to Angerstein Wharf, Greenwich. Its long survival was due in a considerable degree to the devotion of its driver, Alf Peacock.

The last fling of British wagon building was a Sentinel order in 1949 for a hundred six wheelers for the Argentine Government's Patagonian coal mines. An attempt to persuade the National Coal Board to adopt Sentinel steam dumpers met only with official indifference and opposition.

Two further attempts to promote road steam for goods transport have been the Pritchard experiment in Australia, which produced a cleverly designed and workable vehicle, but scarcely a flicker of commercial interest, and Lear's work in America which aims at producing an advanced and pollution-free car or truck. The strong and growing American resistance movement to atmospheric pollution and America's worsening oil supply position have meant that he has had a sympathetic hearing and no overt opposition, but it is too early to judge the real impact of his work.

148 During the Suez fuel crisis in 1956 a few steam wagons were recommissioned. Here, S type Sentinel No 9277 of 1937 (then owned by N Nicholls, Maidstone) was working on contract hire to Fremlins, the Maidstone brewers. Until 1947 it had been owned by the London & Rochester Trading Co Ltd. Painted white, it was used for delivering their stock bricks from North Kent to London.

149 In the words of the late Alfred R Bennett, 'Forgotten, forlorn and for sale' Garrett No 35465, delivered new to Glover & Uglow of Kelly Bray and later used by Cornwall County Council as a tar sprayer, languishes in Jewell's quarry at Sladesbridge, Devon, in the late fifties. Happily it escaped the torch and was restored by Jim Hutchens of Ferndown, Dorset.

150 Typical of the plight of many wagons in the early fifties, the Foden tractors standing in the scrapyard of T T Boughton & Sons at Amersham Common, Bucks, are [in foreground] No 13454 [new to Bucks County Council in 1929 as a tipper] and No 13536, which had already had four owners when Boughtons acquired it in 1938. Both were converted to tractors and both survive.

151 In Caulfield, Victoria, Australia, in the late fifties, two steam engineers, A M & E Pritchard, set about designing their own version of a modern automated steam lorry. Using an oil fired monotube boiler and a pressure of 1000 psi, the lorry was powered by a vertical Vee twin engine with 4ins bore and stroke, mounted behind the cab with cam operated poppet valves for the steam inlets and exhaust was on the Uniflow principle. A prototype vehicle in which the Pritchard boiler, engine and condenser were mounted in an i.c. chassis performed convincingly but was ignored by operators.

152 The problem of 'smog' accentuated by exhaust fumes led the State of California to appoint a commission in 1969 to examine the possibilities of steam as an alternative to the internal combustion engine. Three steam buses were evolved by manufacturers and submitted for evaluation in April 1972. All used standard buses originally designed to be powered by GMC 6771 diesels. Steam Power & Systems of San Diego used a six cylinder double acting compound engine, Lear Steam Motors Corporation of Reno, Nevada, fitted a 75,000 rpm turbine at 1500 psi and Brobeck of Berkeley used a three cylinder compound engine in which a 3½ins diameter high pressure was matched to two 4¾ins diameter low pressures. All used flash steam generators. To date, mass production of any of the designs does not appear to be imminent.

153 The old DG has the last world. Sentinel No 8122 is still spraying in North Wales, in the ownership of Lloyd Jones Bros of Ruthin - the last working wagon in Britain. Sad to relate, whilst this work was being written Murray Lloyd Jones, one of the partners in the owning firm, died. This picture of it was taken in July 1962 by the late Bill Hughes when it belonged to its previous owners, Robert Bridson & Son of Neston, Wirral.

INDEX OF PHOTOGRAPHS

A
Allchin 105, 120A
Atkinson 89A, 100, 142
Aveling & Porter 50, 77

B
Beach 14
Bollée 1, 2
Bretherton & Bryan 39
Brightmore 49
Burrell 62, 62A, 95

C
Carmont
Centre Steer
Clarkson 15, 71, 71A, 99A
Clayton & Shuttleworth 69, 70, 97, 98
Coulthard 18, 32

D
Darracq-Serpollet 35
De Dion Bouton 7, 22

E
Ellis 12

F
Foden 24, 28, 41, 58, 66, 68, 78, 84, 85, 86, 92, 93, 104, 108, 117, 120, 120A, 121, 122, 126, 136 141, 146, 150
Foster 89
Fowler 111
Freibahn Gesellschaft 47

G
Garrett 43, 63, 75, 76, 81, 87, 94, 96, 101, 103, 106, 106A, 110, 115, 134, 149
Gillett 30

H
Henschel 127A, 127B
Hercules 55, 55A
Hindley 38
Howard 33

I
Irgens 20

L
Lear 152
Le Blant 5, 6
Leyland 11, 48, 59, 85A, 143
Lifu 13, 16
Lomax 27
Londonderry 53

M
Manchester 49
Mann 10, 42, 64, 67, 73, 74, 80, 90, 109
Martyn 21
Merryweather 102

N
National - see Clarkson

P
Parmiter 3
Pritchard 151
Purrey 23

R
Robertson 56
Robey 79, 88, 91

S
Sachsenburg 145
Saint Pancras 51
Savage 61
Scotte 4
Sentinel 45, 60, 60A, 69A, 72, 83, 83A 99, 107, 116, 118, 119, 123, 124, 125, 128, 129, 130, 131, 132, 135, 137, 138, 144, 147, 148
Serpollet 8, 34
Sheppee 37, 65
Simpson & Bodman 29
SM 52
Stewart Thornycroft 40
Stoltz 31
Straker 54

T
Tasker 44
Thornycroft 17, 25, 40, 50
Toward 19
Turner 36

W
Wallis & Steevens 57A, 57B
Weidnecht 9

Y
Yorkshire 46, 112, 113, 133

Illustrations 2, 4, 5, 7, 8, 9, 14, 15, 16, 18, 19, 20, 21, 22, 23, 27, 28, 29, 30, 31, 34, 36, 40, 51, 52, 53, 56 and 60A are from the archives of the National Motor Museum Library, Beaulieu, Hants.

Illustrations 10, 12, 13, 32, 38, 39, 45, 62 and 85 are from the archives of the Museum of English Rural Life, Reading, Berks.

First edition copyright Marshall, Harris & Baldwin 1979.
ISBN 0 906116 10 4
All rights reserved. No part of this publication may be reproduced or transmitted in any form or by any means, technical or mechanical, including photocopying, recording or by any information storage and retrieval system without the permission of the publishers.

Published by: Marshall Harris & Baldwin Ltd.
17 Air Street
London, W.1.

Registered in London 1410311.

Designed by: J R Smith

Printed by: Blackwell's, Oxford.

FODENS LIMITED, SANDBACH, CHESHIRE

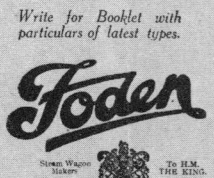

A type for every need in every industry.

The Foden Undertype Wagon ensures power and speed with easy handling.

The Foden Overtype Wagon. Extreme power for heavy loads.

The Foden with its extremely flexible steam engine will haul loads from 6 to 15 tons speedily and economically. For heavy haulage over long or short distances it will give you many years of reliable service.

There is a Foden Wagon for every heavy load in every industry.

Write for Booklet with particulars of latest types.

For the Miller.
For the Brewer. 
For the Haulage Contractor. 
3-way Tipping Wagon. 
General Contractor's Wagon. 
The Foden Overtype Tractor.

Foden
Steam Wagon Makers To H.M. THE KING.

The Mann Express WAGON

4 SPEEDS With the mechanical parts of two only. Totally enclosed Propeller-shaft Transmission running entirely in oil.

Always on its job—fit and ready for an honest day's work. Master of all loads, all roads, and all distances.

Write for our latest booklets on "Service" and the Mann "Overtype" and "Express" Wagons.

All the talk about efficiency in a commercial vehicle boils down to one point—*earning capacity.*

The "Mann" Express Wagon is always earning more—because it works consistently—week in week out. Reliability applied to the "Mann" not only means freedom from mechanical trouble on the road—it means also the ruling out of the need for constant adjustments and frequent overhauls and replacements—which keep a wagon in the shops when it should be earning money on the road.

MANN'S PATENT STEAM CART & WAGON Co. Ltd.
HUNSLET, LEEDS.

London Representative:— J. M. DALLAS, 42 The Avenue, Muswell Hill, N.10.
Spare Parts Depot for London and District:— BECK & POLLITZER, 133/7 Queen Victoria Street, E.C.4.

MAKERS OF THE MANN "OVERTYPE" AND THE MANN "EXPRESS" STEAM WAGONS.

MOTOR COALS

PENRIKYBER NAVIGATION
TYDRAW NAVIGATION } **THE BEST**

CORY BROTHERS & Co., Ltd.,
Colliery Owners, Cory's Buildings, 59, St. Mary Axe, London, E.C.3.

Telegrams: Cory, Stock, London.   Telephones: Avenue 3289 and 1285.
Head Office: Cory's Buildings, Cardiff.

ALL ROADS TO THE GARRETT ARE SMOOTH ROADS

Anything—Anywhere

THIS is the slogan adopted by our customers—the well-known hauliers—Messrs. Murrell's Wharf, of London, S.E.1.

The "Garrett" Rigid Six Wheeler is the Wagon to take anything—anywhere—no matter how rough the road or how steep the hills.

The "Garrett" will handle all types of heavy and bulky loads, in record time—at a cost far below that of any other heavy haulage vehicle with the same load carrying capacity.

Write for our new Book No. B.655. It is full of interesting facts and figures on heavy transport.

Est. 1778

WORKS LEISTON SUFFOLK

RICHARD GARRETT & SONS LTD
ALDWYCH HOUSE, LONDON, W.C.2

ASSOCIATED WITH AGRICULTURAL & GENERAL ENGINEERS LTD.

Clayton Construction

ENSURES ALL THAT IS BEST, AND EMBODIES THE LATEST AND MOST EFFICIENT IMPROVEMENTS IN

Wages have risen and cannot be reduced. The only item in the cost of production on which large savings can be made is the cost of transport.
W. REES JEFFREYS, Roads & Transport Congress.

All this is . . . assured in the **Clayton** WAGON.

STEAM MOTOR WAGON

TELEGRAPHIC ADDRESS: "CLAYTONS, LINCOLN"

CLAYTON & SHUTTLEWORTH, Ltd
. . . LINCOLN . ENGLAND . . .

TELEPHONE: Nos. 42 and 43 LINC.